MATHEMATICS

Formulae & Definitions

CONTENTS

Published by
O.P. Gupta *for* Ramesh Publishing House
Admin. Office
12-H, New Daryaganj Road, Opp. Officers' Mess,
New Delhi-110002 ✆ 23275224, 23245124
E-mail: info@rameshpublishinghouse.com
For Online Shopping: www.rameshpublishinghouse.com
Showroom
- Balaji Market, Nai Sarak, Delhi-6 ✆ 23282525 📱 9354373464
- 4457, Nai Sarak, Delhi-110006

© *Reserved with the Publisher*

No Part of this book may be reproduced or transmitted in any form or by any means, electronic or mechanical including photocopying, recording or by any transformation storage and retrieval system without written permission from the Publisher.

Indemnification Clause: This book is being sold/distributed subject to the exclusive condition that neither the author nor the publishers, individually or collectively, shall be responsible to indemnify the buyer/user/possessor of this book beyond the selling price of this book for any reason under any circumstances. If you do not agree to it, please do not buy/accept/use/possess this book.

Book Code: R-1009
ISBN: 978-81-7812-525-1
27th Edition: October 2025
Price: ₹ 80

Printed at: Deepak Offset, Delhi

R. GUPTA'S®

MATHEMATICS

Formulae & Definitions

by

Ramanand Thakur

Ramesh Publishing House, New Delhi

1

SETS, RELATIONS, FUNCTIONS & BINARY OPERATION

OPERATIONS ON SETS

1. **Symbols:** Some commonly used symbols are as follows:

Symbols	*Meaning*
$\Rightarrow$	implies
$\in$	belongs to
$A \subset B$	A is a subset of B
$\Leftrightarrow$	implies and is implied by
$\notin$	does not belong to
s.t.	such that
$\forall$	for every
$\exists$	there exists
iff	if and only if

&	and
a/b	a is a divisor of b
N	set of natural numbers
I	set of integers
R	set of real numbers
C	set of complex numbers
Q	set of rational numbers

2. **Sets:** A set (class, aggregate, ensemble) S is a well-defined collection of objects, or symbols, called *elements* or *members* of the set.

3. **Representation of Sets**

   (a) Roaster Method (Listing Method): We list all the elements and enclose them in curly brackets; e.g.,

   (i) {2, 3, 5, 7} is the set of prime number less than 10.

   (ii) {a, e, i, o, u} is a set of vowels in English alphabet.

   (iii) {1, 2, 3, 4, ...} is the set of natural numbers.

   (b) Set Builder Method (Property form): The set of all elements of *n*, which satisfy a given property *p* (say) is represented by $\{x : p(x)\}$ or $\{x \mid p(x)\}$ where the symbols ':' or '|' stands for 'such that' and $p(x)$ means x has the property p, e.g.,

$B = \{x : x = 2n, n \in R\}$ or $\{x \mid x = 2n, n \in R\}$

$= \{2, 4, 6, 8, ...\}$

4. **Number Sets:** We give below number sets:

(i) The set of natural numbers
$N = \{1, 2, 3, 4, ...\}$

(ii) The set of integers
Z or $I = \{....., -3, -2, -1, 0, 1, 2, 3, ...\}$

(iii) The set of positive integers
Z^+ or $I^+ = \{1, 2, 3, ...\} = N$

(iv) The set of negative integers
Z^- or $I^- = \{-1, -2, -3, ...\}$

(v) The set of non-negative integers
$W \equiv$ the set of whole nos. $= \{0, 1, 2, 3, ...\}$

(vi) The set of non-zero integers
$Z_0 = \{\pm 1, \pm 2, \pm 3, ...\}$

(vii) The set of all real numbers is denoted by R.

(viii) The set of all irrational numbers $R - Q$.
A number which is real but not rational is called an irrational number.

e.g. π, e, $\sqrt{2}$, $\sqrt{3}$, log 2, etc.

5. **Subset:** If A and B are two sets such that every element of A is also element of B i.e.,

$x \in A \Rightarrow x \in B$, then we say that A is a subset of B or A is contained in B, it is denoted by $A \subseteq B$, some authors write it as $A \subset B$.

6. **Proper and Improper Subsets:** The null set ϕ is subset of every set and every set is subset of itself, i.e., $\phi \subseteq A$ and $A \subseteq A$ for every set A. They are called improper subsets of A. Thus every non-empty set has two subsets, i.e., A has two improper subset iff it is non-empty. All other subsets of A are called its proper subsets.

   Thus if $A \subseteq B$, $A \neq B$, $A \neq \phi$, then A is said to be proper subset of B.

7. **Number of Subset of a finite set:** If set A has n elements, then A has 2^n subsets.

8. **Power Set:** The family (set) of all subsets of a Set A is called the power set of A. It is denoted by P(A).

   Thus, $n(A) = p \Rightarrow n(P(A)) = 2^p$.

Some Important Deductions:

(i) $A \subseteq A, \forall A$

(ii) $\phi \subseteq A, \forall A$

(iii) $A \subseteq U$ (the universal set), $\forall$ A in U

(iv) $A = B \Leftrightarrow A \subseteq B. B \subseteq A.$

9. **Theorem:** If a finite set S has n elements, then the proper set of S has 2^n elements.

Proof: Let a set S contains n elements. The number of the subsets which have no elements at all = ${}^nC_0 = 1$. The null set is a subset of every set.

The no. of those subsets of S each of which has exactly one element = no. of elements in the set $n = {}^nC_1$.

The no. of those subsets of S each of which has exactly two elements out of n elements of S = ${}^nC_2$.

The no. of those subsets of S each of which has exactly r elements out of n elements of S = ${}^nC_r$.

Similarly, the no. of those subsets of which has exactly n elements of $S = {}^nC_n = 1$.

Therefore, total number of subsets of S

$$= {}^nC_0 + {}^nC_1 + {}^nC_2 + \ldots.. + {}^nC_r + \ldots.. + {}^nC_n$$
$$= 1 + {}^nC_1 + {}^nC_2 + \ldots.. + {}^nC_r + \ldots.. + 1$$
$$= (1 + 1)^n = 2^n$$

Hence, if S is a finite set of order n, then the power set of S has 2^n elements. We can say that a set with three elements has 2^3, i.e., 8 subsets.

10. Union of two sets: The union of sets A and B, written $A \cup B$ is the set of all elements that are in A or in B or in A and B both.

Thus, $x \in A \cup B \Leftrightarrow x \in A$ or $x \in B$, and

$x \notin A \cup B \Leftrightarrow x \notin A$ or $x \notin B$.

For example, let A = {a, b, c}, B = {c, g, h}

then, A ∪ B = {a, b, c, g, h}

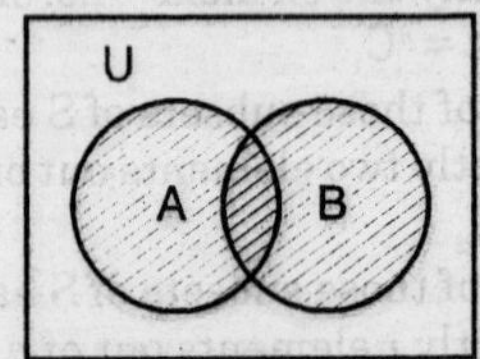

Important Deductions:

(i) A ∪ A = A i.e., the union of set is indempotent.

(ii) A ∪ U = U; where U is the universal set.

(iii) A ∪ ϕ = A; where A is any set.

(iv) A ⊆ A ∪ B; B ⊆ A ∪ B.

(v) A ∪ B = B ∪ A (Commutative law)

(vi) Let A and B be two finite sets such that n(A) = p, n (B) = q, then, min.n (A ∪ B) = max. (p, q) and max. n (A ∪ B) = $p + q$.

(vii) (A ∪ B) ∪ C = A ∪ (B ∪ C) (Associative law)

(viii) $A \cup A' = U$ where A is any subset of the universal set U and A' is complement.

Rule to write $A \cup B$ when A and B are both in Roaster form:

(i) Write all the elements of A.

(ii) Write all the elements of B, dropping the element which are in A.

In Venn diagram, the shaded region under the boundary curve represents $A \cup B$.

Union of more than two sets: If $A_1, A_2, A_3 \ldots\ldots A_n$ are the subset of U and $n \in N$, then the set

$$A_1 \cup A_2 \cup A_3 \cup \ldots\ldots A_n = \bigcup_{i=1}^{n} A_i$$

consists of the members of U which belongs to at least of the subsets of A_i.

For example: if $A = \{a, b, c, d\}$, $B = \{c, d, e, f\}$
$C = \{a, c, e, g\}$, $D = \{b, c, d, g\}$

$A \cup B \cup C \cup D = \{a, b, c, d, e, f, g\}$

11. Intersection of two sets: The intersection of two sets A and B is the set of elements which are common to both A and B and is denoted by $A \cap B$.

This is read as 'the intersection of A and B' or simply 'A intersection B'.

Thus,

$$x \in A \cap B \Leftrightarrow x \in A \text{ and } x \in B.$$

$$\therefore \quad A \cap B = \{x : x \in A \text{ and } x \in B\}.$$

Also, $x \notin A \cap B \Leftrightarrow x \notin A \text{ or } x \notin B$.

For example: If $A = \{a, c, d, e, g, i, r, o, t, u\}$ and $B = \{a, e, i, o, u\}$

then, $A \cap B = \{a, e, i, o, u\}$

If A and B are disjoint sets (i.e., if A and B have no element in common), then, $A \cap B = \phi$.

In Venn diagram, the shaded region under the thick boundary curve represents $A \cap B$.

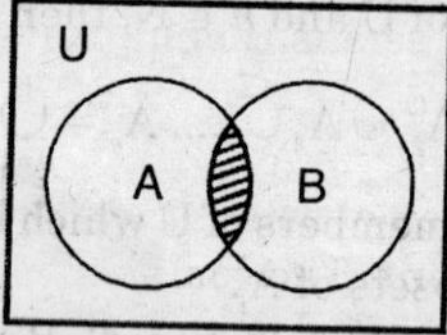

Intersection of more than two sets: If $A_1, A_2, A_3 \ldots\ldots A_n$ are the subset of $\cup$ and $n \in N$, then the set

$$A_1 \cap A_2 \cap A_3 \cap A_4 \cap \ldots\ldots \cap A_n = \bigcap_{i=1}^{n} A_i$$

$$= \{x : x \in A_i \ \forall \ i\text{'s}\}$$

Important Deductions:

(i) $A \cap \phi = \phi$

(ii) $A \cap U = A$

(iii) $A \cap A = A$; i.e., intersection of sets is indempotent.

(iv) $A \cap B \subseteq A$

(v) $A \cap B \subseteq B$

(vi) $A \cap B \subseteq A \cup B$

(vii) $A \subseteq B \Rightarrow A \cap B = A$

(viii) Let A and B be finite sets such that $n(A) = p$, $n(B) = q$.
then, min. $n(A \cap B) = 0$,
max. $n(A \cap B) = \min.(p, q)$

(ix) $A \cap B = B \cap A$ i.e., the intersection of sets is commutative

(x) $A \cap A' = \phi$

(xi) $(A \cap B) \cap C = A \cap (B \cap C)$: i.e., the intersection of sets is associative.

12. Complement: The complement of a set A is the set of all elements of the universal set U which do not belong to A and is denoted by A' (or A^c).

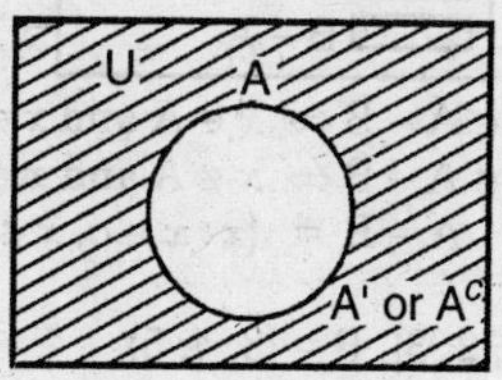

For Example: If $A = \{2, 4, 6, ...\}$
and $U = N = \{1, 2, 3, ...\}$, then
$A' = \{1, 3, 5, 7, ...\}$
also, $(A')' = \{2, 4, 6, ...\} = A.$

Important Deductions:

(i) $A \cap A' = \phi$
(ii) $A \cup A' = U$
(iii) $U' = \phi;\ \phi' = U$
(iv) $(A')' = A$
(v) $A - B = A \cap B'$

13. Difference of Sets: Difference of two sets A and B, denoted by A – B, is the set of all elements of A which are not in B.

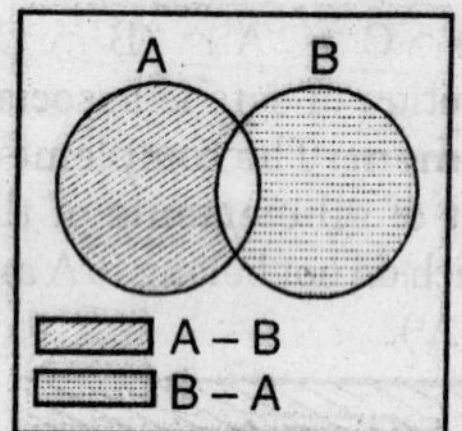

Thus, $x \in A - B \Leftrightarrow x \in A$ and $x \notin B$
and so $x \notin A - B \Leftrightarrow x \notin A$ and $x \in B$
Also, $A - B = \{x : x \in a, x \notin B\}$
For example,

(i) $A = \{1, 2, 3\}$. $B = \{3, 4, 5\}$

Then, $A - B = \{1, 2\}$
$B - A = \{4, 5\}$

(ii) $R - Q$ is the set of all irrational numbers.

Important Deductions:

(i) $A - B \subseteq A$, $B - A \subseteq B$

(ii) $A \subseteq B \Leftrightarrow A - B = \phi$

(iii) $A - B \neq B - A$

(iv) $A - B = A - (A \cap B)$

(v) $A - B$, $B - A$, $A \cap B$ are pair wise disjoint.

14. Symmetric Difference: It is the set of all those elements which are only in A or only in B, i.e., $(A - B) \cup (B - A)$ is called the symmetric difference of A and B. It is usually denoted by $A \Delta B$.

Also, $$A \Delta B = (A \cup B) - (A \cap B).$$

For example, if $A = \{a, e, i, o, u\}$, $B = \{b, d, e, i\}$

then,
$$\begin{aligned} A \Delta B &= (A \cup B) - (A \cap B) \\ &= \{a, b, d, e, i, o, u\} - \{e, i\} \\ &= \{a, b, d, o, u\} \end{aligned}$$

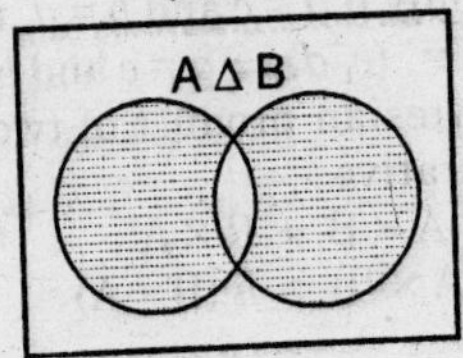

15. Cartesian Product of two sets: Cartesian product of sets A and B denoted by $A \times B$, is the set of all ordered pairs of which first coordinates are elements of the set A and second coordinates, the elements of set B.

Symbolically

$$A \times B = \{(a, b) \mid a \in A \text{ and } b \in B\}$$

$$B \times A = \{(b, a) \mid b \in B \text{ and } a \in A\}$$ is the cartesian product of B and A.

For example:

Let $A = \{a, b\}, B = \{1, 2, 3\}$

Then, $A \times B = \{(a, 1), (a, 2), (a, 3), (b, 1), (b, 2), (b, 3)\}$

Important Deductions:

(i) In a ordered pair, the order of occurrence of members is of prime importance. Ordered pair $(a, 1)$ is not the same as ordered pair $(1, a)$ i.e., $(a, 1) \neq (1, a)$.

(ii) Two ordered pairs (a, b) and (c, d) are equal if and only if $a = c$ and $b = d$, i.e.,

$$(a, b) = (c, d) \Leftrightarrow a = c \text{ and } b = d.$$

(iii) The cartesian product of two sets is not commutative

i.e., $A \times B \neq B \times A$

But, $n(A \times B) = n(B \times A)$

(iv) If A has p elements and B has q elements, then $A \times B$ has pq elements i.e., if $n(A) = p, n(B) = q \Rightarrow n(A \times B) = pq$.

(v) If $A = \phi$ or $B = \phi$, then $A \times B = \phi$

(vi) $B \subset C \Leftrightarrow A \times B \subset A \times C$

16. Some Important Laws:

(i) **Commutative laws**

(a) $A \cup B = B \cup A$

(b) $A \cap B = B \cap A$

(c) $A - B \neq B - A$

(d) $A \Delta B = B \Delta A$

(e) $A \times B \neq B \times A$

(ii) **De-Morgan's law**

(a) $A - (B \cup C) = (A - B) \cap (A - C)$

(b) $A - (B \cap C) = (A - B) \cup (A - C)$

(c) $(A \cup B)' = A' \cap B'$

(d) $(A \cup B)' = A' \cap B'$

(iii) **Distributive law**

(a) $A \cup (B \cap C) = (A \cup B) \cap (A \cup C)$

(b) $A \cap (B \cup C) = (A \cap B) \cup (A \cap C)$

(c) $A \times (B \cap C) = (A \times B) \cap (A \times C)$

(d) $A \times (B \cup C) = (A \times B) \cup (A \times C)$

(e) $A \times (B - C) = A \times B - A \times C$

(iv) **Associative law**

(a) $(A \cup B) \cup C = A \cup (B \cup C)$

(b) $(A \cap B) \cap C = A \cap (B \cap C)$

(c) $(A \Delta B) \Delta C = A \Delta (B \Delta C)$

(d) $(A - B) - C \neq A - (B - C)$

(e) $(A \times B) \times C \neq A \times (B \times C)$

17. Some more results: Let A, B and C be finite sets and U be the finite universal set, then

(a) $n(A \cup B) = n(A) + n(B) - n(A \cap B)$

(b) $n(A \cup B) = n(A) + n(B)$; [if A and B are disjoint set]

(c) $n(A - B) = n(A) - n(A \cap B)$

i.e., $n(A) = n(A - B) + n(A \cap B)$

(d) $n(A \Delta B) = n[(A - B) \cup (B - A)]$

$= n[(A \cup A') \cup (A' \cap B)]$

$= n(A) + n(B) - 2n(A \cap B)$

(e) $n(A' \cap B') = n$ (neither A nor B)

$= n(A \cup B)'$

$= n(U) - n(A \cup B)$

(f) $n(A' \cup B') = n(A \cap B)'$

$= n(U) - n(A \cap B)$

(g) $n(A \cup B \cup C) = n(A) + n(B) + n(C) - n(A \cap B) - n(B \cap C) - n(A \cap C) + n(A \cap B \cap C)$

(h) n (set of elements in exactly two of the sets A, B, C)

$= n(A \cap B) + n(B \cap C) + n(C \cap A) - 3n(A \cap B \cap C)$

(*i*) n (set of elements which are in at least two of the sets A, B, C)

$$= n(A \cap B) + n(A \cap C) + n(B \cap C) - 2n(A \cap B \cap C)$$

(*j*) n (set of elements which are in exactly one of the sets A, B, C)

$$= n(A) + n(B) + n(C) - 2n(A \cap B) - 2n(B \cap C) - 2n(A \cap C) + 3n(A \cap C)$$

(*k*) If A, B, C are three sets, then the distributive law can be easily illustrated by Venn diagrams shown below:

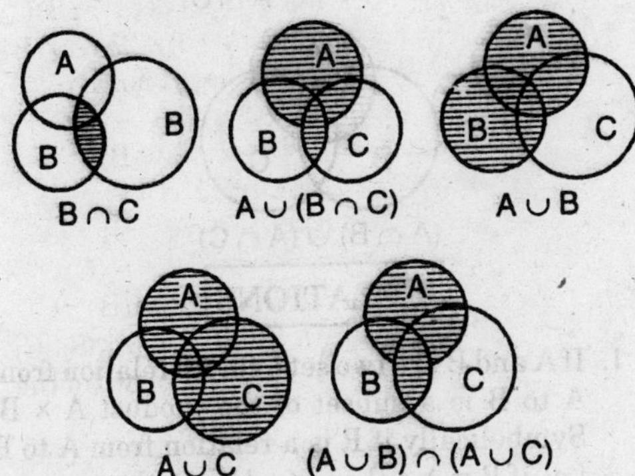

B ∩ C A ∪ (B ∩ C) A ∪ B

A ∪ C (A ∪ B) ∩ (A ∪ C)

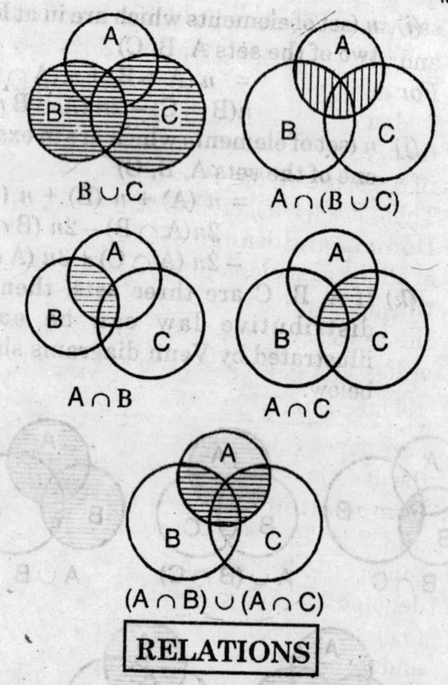

RELATIONS

1. If A and B are two sets, then a relation from A to B is a subset of the product A × B. Symbolically if R is a relation from A to B *i.e.*, if $R \subseteq A \times B$ and $(a, b) \in R$, then we can

say that a is related to b by the relation R and write it as aRb. If $(a, b) \notin$ R then, $a \not R\, b$.

For example:

Let A = {1, 2, 3, 4, 5}, B = {1, 3}

then, we set a relation from A to B as: a R b iff $a \leq b$; $a \in$ A, $b \in$ B

Then, R = {(1, 1), (1, 3), (2, 3)} ⊂ A × B

2. **Domain and Range of relation:** The set of all the first elements of the ordered pairs which belong to R is called the *domain* and the set of all the second elements of that ordered pairs is called the *range*.

Thus, Dom. R ⊆ A; Range R ⊆ B.

For example, in the relation {(6, 8), (3, 7), (1, 2)} the domain is (6, 3, 1) and range is (8, 7, 2).

3. **Composition of Relations:** If R ⊆ A × B and S ⊆ B × C be two relations then composition of the relations R and S is denoted by SoR ⊆ A × C. It is defined by $(a,c) \in$ (S o R) iff $\exists\, b \in$ B such that $(a, b) \in$ R and $(b, c) \in$ S.

For Example,

Let A = {1, 2, 3}, B = {a, b, c, d},

C = {α, β, γ}

R (⊆ A × B) = {(1, a), (1, c), (2, d)}

$S (\subseteq B \times C) = \{(a, \alpha), (a, \gamma), (c, \beta)\}$

Then, $S \, o \, R (\subseteq A \times C) = \{(1, \alpha), (1, \gamma), (1, \beta)\}$

Note: One should be careful in computing the relation R *o* S. Actually R *o* S starts with S and S *o* R starts will R. In general $S \, o \, R \neq R \, o \, S$.

4. **Inverse Relation:** If R be a relation from A set to *a* set B, then the relation R^{-1} from set B to the set A is defined as the inverse relation R.

   Symbolically, $R^{-1} = \{(b, a) : (a, b) \in R\}$

   Hence, it is clear that

   (i) $a \, R \, b \Leftrightarrow b \, R^{-1} \, a$

   (ii) dom. R^{-1} = range R and range R^{-1} = dom. R

   (iii) $(R^{-1})^{-1} = R$.

5. **Void relation in a Set:** Consider the set $A \times A$, then every subset of $A \times A$ is a relation in A. Again the null set ϕ is a subset of $A \times A$, therefore, the null set ϕ is also a relation in A. This relation is called the void relation in A.

   For any ordered pair (a, b) with $a \in A$ and $b \in A$ we have $a \not{R} \, b$ i.e., $(a, b) \notin R$.

   For example, let A = (2, 3, 5) and let R be the relation 'divides' then, R is a void relation since $R = \phi \subset A \times A$.

6. **Reflexive Relation:** R is a reflexive relation if $(a, a) \in R, \forall a \in R$. It should be noted that if $\exists$ any $a \in A$ such that $a \not R a$, then R is not reflexive.
For example, let $A = \{1, 2, 3\}$ and $R = \{(1, 1), (1, 3)\}$, then R is not reflexive since $3 \in A$ but $(3, 3) \in R$.
7. **Symmetric Relation:** R is called a symmetric relation on A if $(x, y) \in R \Rightarrow (y, x) \in R$, i.e., if x is R-related to y, then y is also R-related to x. It should be noted that R is symmetric iff $R^{-1} = R$.
8. **Anti-Symmetric Relation:** R is called anti-symmetric relation if $(a, b) \in R$ and $(b, a) \in R \Rightarrow a = b$.
Thus, if $a \neq b$, then a may be related to b or b may be related to a, but never both.
For example, let N be the set of natural numbers. A relation $R \subseteq N \times N$ is defined by:

$$x \, R \, y \text{ iff } x \text{ divides } y \text{ (i.e., } x/y)$$

Then, $x R y, y R x \Rightarrow x$ divides y, y divides x

$$\Rightarrow x = y$$

It should be noted that the set $\{(a, a) : a \in A\} = D$ is called the diagonal line of $A \times A$.

Then "the relation R in A is antisymmetric iff $R \cap R^{-1} \subseteq D$".

9. **Transitive Relations:** A relation R in a set A is said to be transitive if a R b and b R $c \Rightarrow a$ R c i.e., if $(a, b) \Rightarrow$ R and $(b, c) \in$ R then $(a, c) \in R\ \forall\ a, b, c \in A$.

It should be noted that if a R b, then it is not necessary that b must be related to some element, i.e., if L.H.S. of the implication (*) does not hold, the relation is automatically transitive.

For example, consider the set A = {1, 2, 3} and the relations

$R_1 = \{(1, 2), (1, 3)\}$; $R_2 = \{(1, 2)\}$; $R_3 = \{(1, 1)\}$
$R_4 = \{(1, 2)\ (2, 1), (1, 1)\}$

Then R_1, R_2 and R_3 are transitive while R_4 is not transitive since in R_4 $(2, 1) \in R_4$, $(1, 2) \in R_4$ but $(2, 2) \notin R_4$.

For another example, Let A be the set of all line in a plane R be the relation in A defined by "is parallel to". Then if line a is parallel to line b and line b is parallel to line c, then a is parallel to c. Here, R is transitive.

10. **Identity Relation:** R is an identity relation if $(a, b) \in$ R iff $a = b$. In other words, every element of A is related to only itself.

It is interesting to not that every identity relation is reflexive but every reflexive relation need not be an identity relation.

It should be noted that identity relation is reflexive, symmetric and transitive.

For example,

on the set $= \{1, 2, 3\}$

$R = \{(1, 1), (2, 2), (3, 3)\}$ is the identity relation on A.

11. Equivalence Relation: A relation R in a set A is said to be an equivalence relation if R is reflexive, symmetric and transitive.

(i) R is reflexive, i.e., for every $a \in A$, $(a, a) \in R$ i.e., $a \, R \, a \ \forall \ a \in R$.

(ii) R is symmetric i.e., $(a, b) \in R \Rightarrow (b, a) \in R$ i.e., $a \, R \, b \Rightarrow b \, R \, a \ \forall \ a, b \in A$.

(iii) R is transitive i.e., $(a, b) \in R$ and $(b, c) \in R$ implies $(a, c) \in R$ i.e., $a \, R \, b$ and $b \, R \, c \Rightarrow a \, R \, c$, when $a, b, c \in A$.

For example,

let A be the set of all triangles in a plane and let R be defined by "is congruent to".

We observe that

(i) R is reflexive i.e., $a \, R \, a \ \forall \ a \in A$. Since every triangle is congruent to itself.

(ii) R is symmetric i.e., $a \text{ R } b \Rightarrow b \text{ R } a$. Since, if triangle a is congruent to triangle b then, b is congruent to a.

(iii) R is transitive i.e., $a \text{ R } b$ and $b \text{ R } c \Rightarrow a \text{ R } c$. Since, if triangle a is congruent to triangle b and triangle b is congruent to triangle c then, a is congruent to triangle c.

Thus, the relation R defined above is an equivalence relation.

12. **Equivalence classes of an equivalence relation:** Let R be equivalence relation in A ($\neq \phi$). Let $a \in$ A, then the equivalence class of a, denoted by $[a]$ or $\{\bar{a}\}$ is defined as the set of all those points of A which are related to a under the relation R. Thus $[a] = \{x \in A : x \text{ R } a\}$.

Hence, it is easy to see that

(i) $b \in [a] \Rightarrow a \in [b]$

(ii) $b \in [a] \Rightarrow [a] = [b]$

(iii) Two equivalence classes are either disjoint or identical.

For example, we consider a very important equivalence relation.

$x \equiv y \pmod{n}$ iff n divides $(x - y)$, n is a fixed position integer. Consider $n = 5$. Then,

$[0] = \{x : x \equiv 0 \pmod 5\} = \{5p : p \in z\} = \{0, \pm 5, \pm 10, \pm 15, ...\}$

$[1] = \{x : x \equiv 1 \pmod 5\} = \{x : x - 1 = 5k, k \in Z\}$

$= \{5k + 1 : k \in Z\}$

$= \{1, 6, 11, ..., -4, -9, ...\}$

It should be noted that there are only 5 distinct equivalence classes viz. [0], [1], [2], [3] and [4], when $n = 5$.

FUNCTIONS (MAPPING)

A function is "a rule" or a "device" or "a mechanism" which defines some association or correspondence or relationship between the elements of two sets.

1. **Definition:** Suppose that to each element in a set A there is assigned by some rule, an unique element of a set B. Such rules are called functions or mappings. If we let f denote these rules, then we write $f : A \rightarrow B$ or $A \xrightarrow{f} B$.

   Which reads "f is a function of A into B".

   It is interesting to note that every rule works as a relation but not necessarily as a function.

An Alternative Definition: Let $R \subseteq A \times B$ i.e., R is a relation from A to B. Then R is a function if

(i) dom. R = A.

(ii) $(a, b), (a, c) \in R \Rightarrow b = c$

Terminology: Let $f : A \to B$ be a mapping. Then,

(i) A is called the domain of f.

(ii) B is called the co-domain of f.

(iii) If $a \in A$, then the element in B which is assigned to a, is called the image of a and is denoted by $f(a)$. Thus if $f(a) = b$, b is called the image of a and a is called a pre-image of b.

(**Note:** image of an element in the domain A uniquely exists but pre-image of an element of co-domain may or may not exist in A and if exist, it may not be unique).

(iv) Set of all images is called the range of f. Thus,

Range $f = \{f(a) : a \in A\} \subseteq B$.

Range of f is also written as $f(A)$.

Important Deductions: In a function $f : A \to B$

(i) Each element of the set A must be associated with unique element of B.

(ii) Two or more elements of set A may be associated with the same element of the set B.

(iii) There may be some elements of B which are not assigned to any element of the set A.

2. **Equal Functions:** If f and g are functions defined on the same domain A and if $f(a) = g(a)$ for every $a \in A$, then $f = g$.

For example,

Let $f : \{1, 2\} \rightarrow \{1, 2, 3, 4\}$;
$f = \{(1, 1), (2, 3)\}$ and
$g : \{1, 2\} \rightarrow \{1, 3, 4, 5\}$;
$g = \{(1, 1), (2, 3)\}$

Then, since dom. f = dom. g and $f(a) = g(a)$ for all a in the domain, $f = g$.

3. **Types of Functions:**

(i) ***Constant Function :*** Let $f : A \rightarrow B$. If $f(a) = b$ for all $a \in A$, then f is called a constant function. Thus, f is called a constant function if range f consists of only one element.

(ii) ***Into function:*** Let $f : A \rightarrow B$, if there exists even a single element in B having no pre-image at all in A, then such a function is said to be an into function.

For example,

Let $A = \{1, 2, 3\}$ and $B = \{1, 4, 9, 16\}$

Let $f: A \to B; f(x) = x^2 \ \forall \ x \in A$

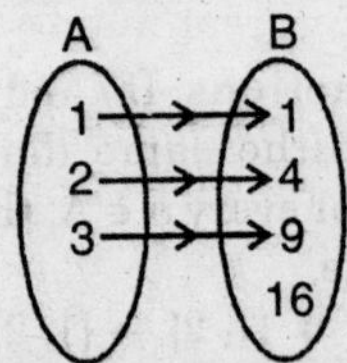

From the figure, it is clear that there exists an element, namely $16 \to B$, which has no pre-image in A. So f is an into function.

(iii) ***Onto function (Surjection):*** A function $f: A \to B$ is said to be an onto function if each element of B is the f image of at least one element of A. An onto function is also called a surjective mapping.

In onto mapping, we observe that

$\{f(x)\} = B \ \forall \ x \in A$

Thus, f is onto iff $f(A) = B$,

i.e., range = co-domain

[**Note:** A function which is not onto, is called an into function].

For example, consider $f = R \to R : f(x) = x^2$. Then f is not an onto function since the negative numbers do not appear in the range of f i.e., no negative number is the square of a real number. So f is an into function.

Important Deduction:

(*i*) In some cases, the following method to test the onto property is useful. Consider $a \in A$, $b \in B$ and $f(a) = b$. Find a in terms of b. If a exists in A for every $b \in B$, f is onto otherwise into.

Illustration: Consider the function

$$f : R \to R : f(x) = \frac{x+1}{x-2}$$

Clearly f is not a mapping in the domain R as $f(2) = \frac{3}{0}$ does not exist in R . So dom. $f = R - \{2\}$. On this domain, consider $a \in R - \{2\}$, $b \in R$,

$$f(a) = b \Rightarrow b = (a+1)/(a-2)$$

$$\Rightarrow ab - 2b = a + 1$$

$$\Rightarrow a = (2b+1)/(b-1)$$

Thus $a \notin (\text{R} - \{-2\})$ for $b = 1$. Thus 1 is not image of any member of the domain so f is into. This method is also useful to find the range. In the above example $f = \text{R} - \{1\}$.

(iv) ***Constant Function:*** Let $f : \text{A} \to \text{B}$. If $f(a) = b$ for all $a \in \text{A}$, then f is called a constant function. Thus f is called a constant function if range f consists of only one element.

(v) ***One-One Function (Injection):*** Let $f : \text{A} \to \text{B}$. Then f is called a one-one function if no two different elements in A have the same image i.e., different elements in A have different elements in B. Symbolically, f is one-one if

$$f(a) = f(a')$$

$$\Rightarrow \quad a = a'$$

i.e., $a \neq a' \Rightarrow f(a) \neq f(a')$

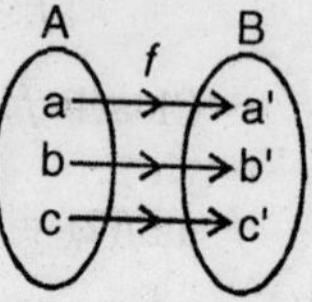

A mapping which is not one-one is called many-one.

For example,

let $f : \text{R} \to \text{R}$:

$f(x) = x^2$. Since, $f(-1) = (1) = 1$, f is not one-one so it is many-one.

Important Deductions:

(i) If any line parallel to x-axis in the graph of the function on cartesian plane intersect the graph at two or more points, f must be many-one. For example, consider the function $y = f(x) = x^2$

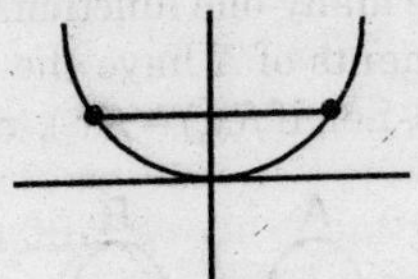

Therefore f is many-one.

(ii) Let $f(x) = f(y)$. If solution of this equation has solution other than $x = y$, f is many-one. For example, let $f : \mathrm{R} \to \mathrm{R}$.

$$f(x) = \frac{x^2}{(x+1)},$$

then $f(x) = f(y)$ gives $\dfrac{x^2}{x+1} = \dfrac{y^2}{y+1}$

$$\Rightarrow x^2y + x^2 = y^2x + y^2$$

$$\Rightarrow xy\,(x-y) + (x-y)\,(x+y) = 0$$

$$\Rightarrow (x-y)(xy+x+y)=0$$

$$\Rightarrow x=y \text{ or } xy+x+y=0$$

$$\Rightarrow x=y \text{ or } y=\frac{-x}{x+1}$$

$\therefore$ f is many-one.

(vi) ***Many-one function:*** $f: A \rightarrow B$ is said to be many-one function, if two or more elements of A have the same f image in B, i.e., if $f(x_1) = f(x_2)$, $x_1 \neq x_2$.

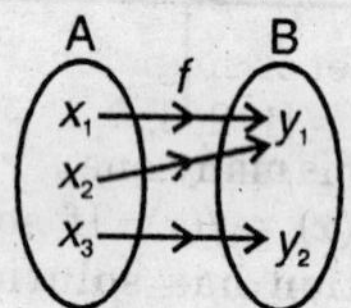

Many-one into mapping:

Let f: A $\rightarrow$ B then

(i) many-one

(ii) into, then f is called a many-one into mapping. In this mapping some elements of B will remain uncovered.

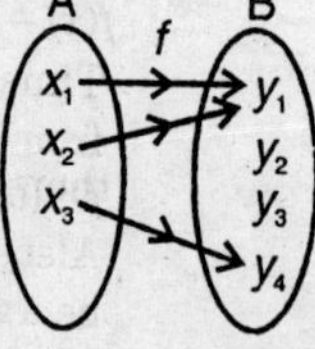

Many-one onto mapping: Let $f: A \to B$, then

(*i*) many-one

(*ii*) onto, then f is called a many-one onto mapping.

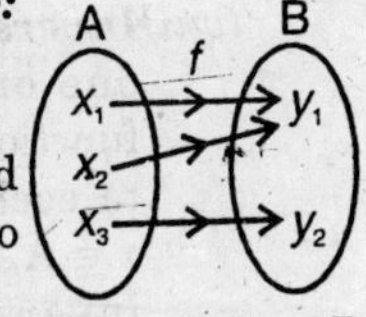

(*vii*) ***Inverse of a function:*** Let $f: A \to B$ and let $b \in B$. Then, the inverse of b, i.e., $f^{-1}(b)$ consists of those elements in A which are mapped onto b, i.e., $f^{-1}(b) = \{x : x \in A, f(x) = b\}$. Therefore, $f^{-1}(b) \subseteq A$. $f^{-1}(b)$ may be a null set or singleton. For example, Let $A = \{1, 2, 3, 4\}$ $B = \{a, b, c\}$ Define, $f = A \to B$ by $f = \{(1, a)\ (2, a)\ (3, b), (4, a)\}$. Then, $f^{-1}(a) = \{1, 2, 4\}$, $f^{-1}(b) = (3)$, $f^{-1}(c) = \phi$

In the same way, if C be a subset of B, f^{-1} (C) is defined by f^{-1} (C)

$$= \{x : x \in A, f(x) \in C\}.$$

Thus, f^{-1} is defined as a rule (relation) from B to A. But then dom. $f^{-1} \subset B$, i.e., $f^{-1}(b)$ may not belong to some $b \in B$. Also, $f^{-1}(b)$ may be unique. So f^{-1} may be a function from B to A.

(viii) ***Inverse Function:*** Let $f: A \to B$ be a one-one onto function. Then the function of $f^{-1}: B \to A$ which associates to each element $y \in B$, the element $x \in A$, such that $f(x) = y$ is called the inverse function of the function $f: A \to B$

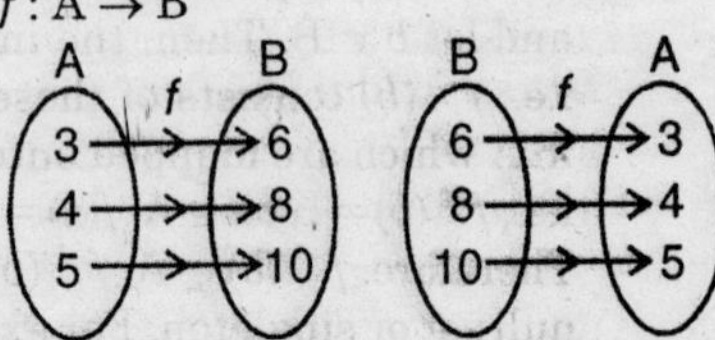

Domain of $f^{-1} = B$ and range of $f^{-1} = A$.

It is easy to see that $f^{-1}: B \to A$ is also one-one onto

$A = \{3, 4, 5\}$, $B = \{6, 8, 10\}$

$f: A \to B, f(x) = 2x, f(3) = 6, f(4) = 8, f(5) = 10$

$f^{-1}: B \to A, f^{-1}(x) = x/2, f^{-1}(6) = 3, f^{-1}(8) = 4,$
$f^{-1}(10) = 5$

$f = (3, 6), (4, 8), (5, 10)$

$f^{-1} = (6, 3), (8, 4), (10, 5).$

The function f^{-1}; $B \to A$ is also one-one onto function.

Uniqueness of inverse function

If $f: A \to B$ is one-one onto function, the inverse function of f is unique.

(*ix*) ***Composition Function or product function:*** Let $f: A \rightarrow B$ and $g: B \rightarrow C$ be two functions. Then the product or composition of the functions f and g, denoted by $g \circ f$, is a mapping of A into C given by $g \circ f: A \rightarrow C$, such that $(g \circ f)(x) = g(f(x)), \forall x \in A$.

Illustration: Let $f: A \rightarrow B$, $f(x) = 2x$; $g: B \rightarrow C$, $g(x) = x^2$, where $A = \{1, 3\}$, $B = \{2, 6\}$, $C = \{4, 36, 12\}$

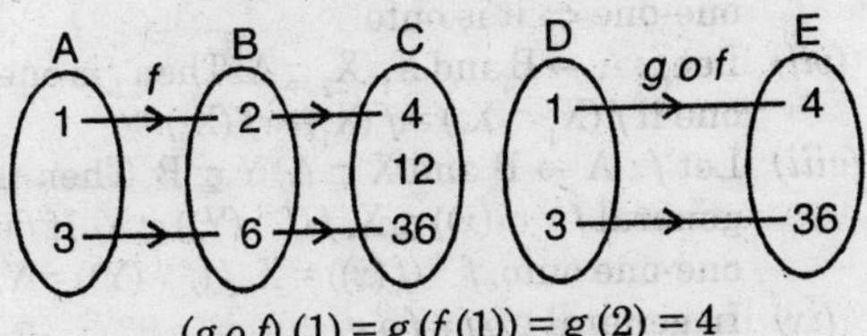

$(g \circ f)(1) = g(f(1)) = g(2) = 4$
$(g \circ f)(3) = g(f(3)) = g(6) = 36$
$(g \circ f)(x) = g(f(x)) = g(2x) = 4x^2$

Thus, $g \circ f: A \rightarrow C, (g \circ f)(x) = 4x^2$
$f = \{(1, 2), (3, 6)\}$
$g = \{(2, 4), (6, 36)\}$
$g \circ f = \{(1, 4), (3, 36)\}$

Important Deductions:

(*i*) If $O(A) = m$, $O(B) = n$, then total number of mappings from A to B is n^m.

(ii) If A and B are finite sets and $O(A) = m$, $O(B) = n$, $m \leq n$. Then number of injection (one-one) from A to B is ${}^nP_m = n!/(n - m)!$

(iii) If $f : A \to B$ is injective (one-one), then $O(A) \leq O(B)$.

(iv) If $f : A \to B$ is surjective (onto), then $O(A) \geq O(B)$.

(v) $f : A \to B$ is bijective (one-one onto), then $O(A) = O(b)$.

(vi) Let $f : A \to B$ and $O(A) = O(B)$. Then f is one-one $\Leftrightarrow$ it is onto.

(vii) Let $f : A \to B$ and $X_1, X_2 \subseteq A$. Then f is one-one if $f(X_1 \cap X_2) = f(X_1) \cap f(X_2)$.

(viii) Let $f : A \to B$ and $X \subseteq A$, $Y \subseteq B$. Then in general $f^{-1}(f(x)) \subseteq X$, $f(f^{-1}(Y)) \subseteq Y$. If f is one-one onto, $f^{-1}(f(x)) = X$, $f(f^{-1}(Y)) = Y$.

(ix) In general $g o f \neq f o g$.

(x) $f : A \to B$, be one-one, onto, then $f^{-1} o f = I_A$ and $f o f^{-1} = I_B$

(xi) $f : A \to B$, $g : B \to C$, $h : C \to D$. Then $(h o g) o f = h o (g o f)$.

(xii) $f : A \to B$, $g : B \to C$ be one-one and onto, then $g o f : A \to C$ is also one-one onto and $(g o f)^{-1} = f^{-1} o g^{-1}$.

(xiii) Let $f : A \to B$, then $I_B o f = f$ and $f o I_B = f$. It should be noted here that $f o I_B$ is not defined since for $(f \circ I_B)(x) = f \circ \{I_B(x)\} = f(x)$, $I_B(x)$ exist when $x \in B$ and $f(x)$ exist when $x \in A$.

(xiv) $f: A \to B$, $g: B \to C$ are both one-one, then $g \circ f: A \to C$ is also one-one but converse is not true i.e., though $g \circ f$ is one-one, both f and g need not be one-one. It should be noted that for $g \circ f$ to be one-one, f must be one-one.

(xv) If $f: A \to B$, $g: B \to C$ are both onto then $g \circ f$ must be onto. However, the converse is not true. But for $g \circ f$ to be onto g must be onto.

(xvi) The domain of the function $(f + g)(x) = f(x) + g(x)$, $(f - g)(x) = f(x) - g(x)$, $(fg)(x) = f(x).\ g(x)$ is given by (dom. f) $\cap$ (dom. g). While domain of the function $(f/g)(x) = f(x)/g(x)$ is given by (dom. f) $\cap$ (dom. g) $- \{x : g(x) = 0\}$.

OPERATION

An operation over a set is a rule which combines any two elements of the set.

Binary Operation: An operation O is called a binary operation on a set A if $a \text{ O } b \in A$, $a, b \in A$.

If O is a binary operation on the set A, then A is said to be closed with respect to the operation O.

For Example:

(*i*) If we consider the composition of multiplication in the set of odd integers, then it is binary composition, since the multiplication of two odd integers is an odd integer i.e.,

$5 \times 3 = 15, \quad 9 \times 5 = 45$

(*ii*) Division is not a Binary operation on N, because $2/3 \in N$. Thus, N is not closed with respect to division operation.

(*iii*) Subtraction is a binary operation on the set of integers I, because $a - b \in I$, for all $a, b \in I$.

Laws of Binary Operation

(*i*) ***Commutative Composition:*** A composition in a set A is used to be commutative if

$$x o y = y o x, \ \forall x, y \in A$$

For example,

$$5 + 4 = 4 + 5$$

$$5 \times 4 = 4 \times 5$$

This shows that addition and multiplication composition in the set of integers are commutative. It should be noted that subtraction is not commutative composition on the set of real numbers as

$$a - b \neq b - a \quad \forall\, a, b \in R$$

e.g., $8 - 2 \neq 2 - 8$

(ii) ***Associative Composition:*** A composition in a set of elements A is said to be association if

$$(x \, o \, y) \, o \, z = x \, o (y \, o \, z), \forall\, x, y, z \in A$$

For example, addition in an associative composition in the set of natural numbers, as we have

$$(5 + 2) + 4 = 5 + (2 + 4) \text{ etc.}$$

In general $(x + y) + z = x + (y + z)$

$$\forall\, x, y, z, \in N$$

Multiplication is an associative composition in the set of integers, since

$$(5 \times 2) \times 4 = 5 \times (2 \times 4)$$

In general $(x \times y) \times z = x \times (y \times z)$

$$\forall\, x, y, z \in I$$

(iii) ***Distributive Composition:*** If *o* and * are two binary composition defined on a set A, then

(a) * is said to be left distributive over *o* if

$(y \, o \, z) * x = (y * x) \, o \, (z * x) \; \forall \; x, y, z \in I$

(b) $*$ is said to be right distributive over o if.

$x * (y \, o \, z) = (x * y) \, o \, (x * z), \forall \; x, y, z, \in A$

If both (a) & (b) hold then we say $*$ is distributive over o.

For example, multiplication is left distributive over addition in the set of real numbers if

$$(b + c) \times a = b \times a + c \times a, \forall \; a, b, c \in R$$

Multiplication is right distributive over addition in the set of real numbers if

$$a \times (b + c) = a \times b + a \times c, \forall \; a, b, c, \in R.$$

(iv) The binary operation O is said to be indempotent on a set of element A, if for every $a \in A$.

$$a \, o \, a = a$$

(v) An element e in a set A is said to be a unit element with respect to the Binary operation o on A if for every $a \in A$.

$$a \, o \, e = e \, o \, a = a$$

(vi) An element b in a set A is said to be the inverse element of an element $a \in A$ with respect to the binary operation o if

$a \, o \, b = b \, o \, a = e$ {If e exists in A}

Operation Table: When the set A being considered has a small number of elements then the result of applying the binary operation *o* to its elements may be represented in the table known as operation table. We write the elements of A in the same order both vertically horizontally. The result *a o b* then appears in the body of table at the intersection of row headed by a and the column headed by *b*.

o	*a*	*b*	*c*
a	*b*	*c*	*b*
b	*a*	*c*	*b*
c	*c*	*b*	*a*

2

COMPLEX NUMBERS

1. **Complex Numbers:** The number $a + ib$, where a and b are real numbers, i.e., "The order pair (a, b), where a and b are real numbers, is called a complex number."

   If $z = a + ib$, then a is called the real part, denoted by $Re\ (z)$ and b is the imaginary part denoted by $Im\ (z)$ of the complex number $Z = a + ib$.

   A complex number $z = x + iy$, $x, y \in R$, is purely real if $y = 0$ i.e., if $Im\ (z) = 0$ and purely imaginary if $x = 0$, i.e., $Re\ (z) = 0$.

2. **Equal complex numbers:**

   Let $z_1 = x_1 + iy_1$ and $z_2 = x_2 + iy_2$

   then, $z_1 = z_2$. i.e., $x_1 + iy_1 = x_2 + iy_2$

   $\Leftrightarrow x_1 = x_2, y_1 = y_2$

It should be noted that a complex number $z = x + iy = 0 \Leftrightarrow x = 0, y = 0$.

3. **Need for Complex Numbers:** Consider the equation $x^2 + 1 = 0$ or $x^2 = -1$. There exists no real number x which may satisfy that equation. Thus, this arises need to extend the system of real numbers to a system in which equations of the above type can be solved. Euler was first to introduce the symbol of i for the positive square root of -1. The number $\sqrt{-1}$ denoted by i (read as iota) is called the imaginary number.

Integral power of i

$$i = \sqrt{-1} \therefore i^2 = -1$$
$$i^3 = i^2.i = (-1) \times i = -i$$
$$i^4 = (i^2)^2 = (-1)^2 = 1$$

A number of the form $x + iy$ where x and y are real numbers and $i = \sqrt{-1}$ is called a complex number. The set of complex number is denoted by the letter C.

If $z = x + iy$ is a complex number, then x is called the real part of z and we write Re $(z) = x$, y is called the imaginary part of z and we write Im $(z) = y$.

If $x = 0$ and $y \neq 0$, the complex numbers reduces to the form iy, which is called a pure imaginary number. If $x \neq 0$ and $y = 0$, then the complex number reduces to form x which is a real number.

4. **Set of Complex Number:** The set of complex numbers is denoted by C,
   Where, $C = \{z : z = a + ib, a, b \in R\}$
   or $C = \{z : z = (a, b), \forall a, b \in R\}$
5. **Zero Complex Number:** If complex number $z = a + ib$, $a, b \in R$ is said to be zero complex number or zero of C if and only if $a = 0$ and $b = 0$.
6. **Negative of a Complex Number:** The complex number $-z = -a - ib$ is called the negative of the complex number $z = a + ib$ and vice-versa.
7. **Equality of two Complex Numbers:** Two complex numbers are said to be equal if and only if their real and imaginary parts are separately equal, i.e., $a + ib = c + id \Leftrightarrow a = c$ and $b = d$.
8. **Algebra of Complex Numbers:**
   (*i*) Addition of two complex numbers: If $z_1 = a + ib$ and and $z_2 = c + id$, $(a, b, c, d \in R)$ are two complex numbers,

then, their sum is $z_1 + z_2$ and is defined as $z_1 + z_2 = (a + ib) + (c + id)$
$$= (a + c) + i(b + d)$$

Hence, the sum of two complex numbers is again a complex number.

(ii) Subtraction of two complex numbers:

If $z_1 = a + ib$ and $z_2 = c + id$ $(a, b, c, d \in R)$.

then, $z_1 - z_2 = z_1 + (-z_2)$; when $-z_2$ is a negative of z_2.
$$= (a + ib) + (-c - id)$$
$$= (a - c) + i(b - d)$$

Hence, the difference of two complex numbers is again a complex number.

(iii) Multiplication of two complex numbers:

Let $z_1 = a + ib$ and $z_2 = c + id$ be two complex numbers, then their product $z_1 z_2$ is defined by
$$z_1 z_2 = (a + ib)(c + id)$$
$$= (ac - bd) + i(bc + ad)$$

Hence, the multiplication of two complex numbers is a complex number.

(iv) Division of two complex numbers: If $z_1 = a + ib$ and $z_2 = c + id$ ($z_2 \neq 0$ i.e., $c \neq 0$, $d \neq 0$) be two complex numbers, then there exists complex number $z = x + iy$ and such that

$z_1 = z.z_2$

Then, complex number $z = (x + iy)$ is called the quotient of z_1 and z_2 and $z = \frac{z_1}{z_2}$.

9. Basic Properties of Complex Numbers:

(a) Properties of addition in C.

(i) ***Addition is closed in C:*** If $z_1 = a + ib$ and $z_2 = c + id$ are two complex numbers, then $z_1 + z_2 = (a + ib)\ (c + id) = (a + c) + i\,(b + d)$ which is a complex number.

(ii) ***Addition is commutative in C:*** If $z_1 = a + ib$ and $z_2 = c + id$ are two complex numbers, then

$$z_1 + z_2 = z_2 + z_1$$

(iii) ***Addition is associative in C:*** If $z_1 = a + ib$, $z_2 = c + id$, $z_3 = e + if$ are three complex numbers, then

$$z_1 + (z_2 + z_3) = (z_1 + z_2) + z_3$$

(iv) ***Existence of additive identity in C:*** If $a + ib$ is complex number, then $(a + ib) + (0 + i0) = (a + 0) + i\,(b + 0)$ $= a + ib$

Thus, the complex number $0 + i0$ is the additive identity in C.

(v) ***Existence of inverse in C :*** For every complex number $a + ib$, there exist a complex number $(-a) + i(-b)$ such that $[a + ib] + [(-a) + i(-b)] = [a - a] + i(b - b) = 0 + i\,0 = 0$ = Additive identity of C.

(b) Properties of multiplication

(i) ***Closure law:*** If z_1 and z_2 are two complex numbers, then $z_1 z_2$ is also a complex number.

(ii) ***Associative law:*** If z_1, z_2, z_3 are any three complex numbers, then $(z_1 z_2) z_3 = z_1 (z_2 z_3)$.

(iii) ***Identity law :*** $1 + i0$ is the multiplication identity of C.

(iv) ***Inverse law:*** For every complex number $z = a + ib$ there exists a complex number $(x + iy)$ such that $(a + ib)(x + iy) = 1 + i0$ = Multiplicative identity of C.

This complex number $x + iy$ is called multiplicative inverse of $(a + ib)$.

Rule to find the multiplicative inverse.

If $z = a + ib$ is a non-zero complex

number, then multiplicative inverse of z

is $\frac{1}{z} = \frac{1}{a+ib} = \frac{a-ib}{a^2+b^2} = \frac{z}{|z|^2}$

(*v*) ***Commutative law:*** If z_1 and z_2 are any two complex numbers, then

$$z_1 z_2 = z_2 z_1$$

(*vi*) ***Distributive law:*** If z_1, z_2, z_3 are any three complex numbers, then

$$z_1 (z_2 + z_3) = z_1 z_2 + z_1 z_3$$

10. Modulus and Argument of a Complex Number: Let $z = x + iy$, then the modulus of z is the positive real number $\sqrt{(x^2+y^2)}$ and is denoted by $|z|$. *i.e.*,

$$|z| = \sqrt{(x^2+y^2)}$$

The argument or amplitude of a complex number $z = x + iy \neq 0$ is any one of the numbers which are solution of the system of equations.

$\cos\theta = \frac{x}{\sqrt{(x^2+y^2)}}$; $\sin\theta = \frac{y}{\sqrt{(x^2+y^2)}}$ is denoted by arg(z) or Arg(z)

The argument of a complex number is not unique. If θ is a value of the argument, then $2n\pi + \theta$, where $n \in I$, are also values of the arguments of z.

Principal value of the argument of the complex numbers z is the argument θ, which satisfy the inequality $-\pi < \theta \le \pi$.

Note that the argument of the complex number 0 is not defined.

11. Properties of Argument:

(i) The argument of the product of any finite number of complex numbers is equal to sum of their arguments.

$$\therefore \arg (z_1.z_2.z_3 \ldots\ldots z_n) = \theta_1 + \theta_2 + \theta_3 + \ldots\ldots + \theta_n$$

$$= \arg z_1 + \arg z_2 + \ldots\ldots + \arg z_n$$

(ii) If z_1 and z_2 are two complex numbers, then $|z_1 + z_2|^2 + |z_1 - z_2|^2$

$$= 2 [\,|z_1|^2 + |z_2|^2]$$

(iii) $\arg \left(\dfrac{z_1}{z_2}\right) = \arg z_1 - \arg z_2$

12. Conjugate of a Complex Number: If $z = x + iy$ is a complex number, then the complex number $x - iy$ denoted by $\bar{z}$ is called the conjugate of the complex number z, *i.e.*, $\bar{z} = x - iy$.

Properties of Conjugate and Moduli:

(*i*) $|z| = \bar{z}$

(*ii*) $|z^2| = z.\bar{z}$

(*iii*) $\overline{(z_1 \pm z_2)} = \overline{z_1} \pm \overline{z_2}$

(*iv*) $|z|^2 = z.\bar{z}$

(*v*) $\overline{(z_1 z_2)} = \bar{z}_1.\bar{z}_2$

(*vi*) $\overline{(z_1 / z_2)} = \bar{z}_1 / \bar{z}_2$

(*vii*) $|z_1 + z_2| \leq |z_1| + |z_2|$

(*viii*) $|z_1 - z_2| \geq |z_1| - |z_2|$

13. **Geometrical Representation of Complex Number:** A complex number $z = x + iy$ can be represented by a point P whose cartesian coordinates are (x, y) referred to the axes OX and OY, called the real and imaginary axes respectively.

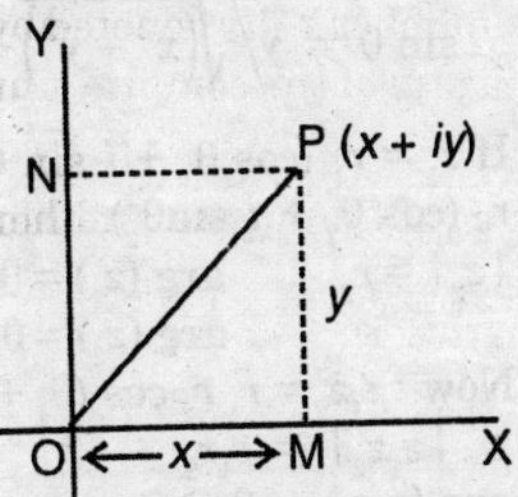

The plane in which we represent the complex numbers is called the Argand plane or Argand diagram or Complex plane or Gaussian plane.

14. Polar Representation of a Complex Number:

(i) The complex number $z = x + iy$, $x, y \in R$ can be represented in polar form as $z = r(\cos\theta + i\sin\theta)$; where $x = r\cos\theta$, $y = r\sin\theta$, $r = \sqrt{(x^2 + y^2)}$ $= |z|$ and θ, which is arg z, is the solution of the equations.

$$\cos\theta = x/\sqrt{(x^2 + y^2)} = \frac{x}{r} \text{ and}$$

$$\sin\theta = y\Big/\sqrt{(x^2+y^2)} = \frac{y}{r}$$

(ii) If $z_1 = r_1(\cos\theta_1 + i\sin\theta_1)$ and $z_2 = r_2(\cos\theta_2 + i\sin\theta_2)$; then $|z_1| = r_1$, $|z_2| = r_2$, $\arg(z_1) = \theta_1$ and $\arg(z_2) = \theta_2$

Now $z_1z_2 = r_1 r_2 \cos(\theta_1 + \theta_2)$

$\therefore\ |z_1z_2| = r_1r_2$

$\arg(z_1z_2) = \theta_1 + \theta_2$

Thus, we have $|z_1z_2| = |z_1|\ |z_2|$ and $\arg(z_1z_2) = \arg z_1 + \arg z_2$.

Also, $(z_1/z_2) = (r_1/r_2)\cos(\theta_1 - \theta_2)$

$\therefore\ |z_1/z_2| = (r_1/r_2)$ and $\log(z_1/z_2) = \theta_1 - \theta_2$

Hence, we have $|z_1/z_2| = |z_1|/|z_2|$ and $\arg(z_1/z_2) = \arg z_1 - \arg z_2$.

15. Cube Roots of Unity:

The cube roots of unity are given by 1,

$$\omega = \frac{-1+\sqrt{3}i}{2},\ \omega^2 = \frac{-1-\sqrt{3}i}{2}$$

Properties of Cube roots of unity:

(i) The sum of the three cube roots of unity is zero *i.e.,* $1 + \omega + \omega^2 = 0$

(ii) The product of the cube roots of unity is one

i.e., $1.\omega.\omega^2 = 1$

i.e., $\omega^3 = 1$

(iii) Each complex cube root of unity is the square of the other.

(iv) Each complex cube root of unity is the reciprocal of the other.

If $\alpha = \frac{-1+\sqrt{3}}{2}$ and

$$\beta = \frac{-1-\sqrt{3}}{2}$$

Then, $\alpha = \frac{1}{\beta}$ and $\beta = \frac{1}{\alpha}$

16. *n*th Roots of Unity:

(i) The n, nth roots of unity are $\alpha_r = \cos(2\pi r/n) + i \sin(2\pi r/n) = e^{(2\pi r/n)i}$, where $r = 0, 1, 2, \ldots, (n-1)$.

(ii) If α is one of the nth root of unity, then $\alpha^n = 1$.

(iii) The sum of n, nth roots of unity is zero, i.e., $1 + \alpha_1 + \alpha_2 + \ldots + \alpha_{n-1} = 0$

(iv) The product of n, nth roots of unity is $(-1)^{n-1}$, i.e., $1.\alpha_1.\alpha_2 \ldots.. \alpha_{n-1} = (-1)^{n-1}$

17. Distance between two points: If z_1, z_2 are two complex numbers then the distance between z_1 and z_2 is $|z_1 - z_2|$.

18. Point dividing a line segment in a given ratio: Let $z_1 = x_1 + iy_1$, $z_2 = x_2 + iy_2$ be the affixes of the points A and B respectively in the argand plane. If λ be a real number $\neq -1$, then there is a unique point C on AB such that AC : CB = λ : 1.

The point C is given by $\dfrac{x_1 + \lambda x_2}{\lambda + 1}, \dfrac{y_1 + \lambda y_2}{\lambda + 1}$

The affix of C is therefore, $(z_1 + \lambda z_2)/(1 + \lambda)$

Remark: The affix of the mid-point of z_1, z_2 is $(z_1 + z_2)/2$.

19. If z_1, z_2, z_3 be the affixes of the vertices of a triangle, the centroid of the triangle has the affix $(z_1 + z_2 + z_3)/3$.

20. Three points z_1, z_2, z_3 are collinear if

$$\begin{vmatrix} z_1 & \bar{z}_1 & 1 \\ z_2 & \bar{z}_2 & 1 \\ z_3 & \bar{z}_3 & 1 \end{vmatrix} = 0.$$

21. De-moivre's Theorem: If n is any integer, then

$$(\cos\theta + i\sin\theta)^n = \cos n\theta + i\sin n\theta$$

$$(\cos\theta + i\sin\theta)^{-n} = \cos n\theta - i\sin n\theta = \cos(-n\theta) + i\sin(-n\theta)$$

$$(\cos\theta - i\sin\theta)^n = \cos n\theta - i\sin n\theta = \cos(-n\theta) + i\sin(-n\theta)$$

22. Some Applications of De-Moivre's theorem

(*i*) *To express cos nθ and sin nθ in powers of cosθ and sinθ:*

$$(\cos n\theta + i\sin n\theta) = (\cos\theta + i\sin\theta)^n = \cos^n\theta + {}^nC_1 \cos^{n-1}\theta . i\sin\theta + {}^nC_2 \cos^{n-2}\theta . i^2\sin^2\theta + \ldots\ldots\ldots$$

Equating real and imaginary parts, we get

$$\cos n\theta = \cos^n\theta - {}^nC_2 \cos^{n-2}\theta + {}^nC_4 \cos^{n-4}\theta \sin^4\theta - \ldots\ldots$$

$$\sin n\theta = {}^nC_1 \cos^{n-1}\theta \sin\theta - {}^nC_3 . \cos^{n-3}\theta \sin^3\theta + \ldots\ldots\ldots$$

(*ii*) *To find roots of a complex number:* Let $z = x + iy$, and we have to find its

nth root, putting z in polar form, we get

$$z = r(\cos\theta + i\sin\theta);\ r = |z| = \sqrt{(x^2 + y^2)},$$

and θ is the principal arg of z. Then considering general value of arg z, we have

$$z = r[\cos(2k\pi + \theta) + i\sin(2k\pi + \theta)], \text{ then}$$

$$z^{1/n} = r^{1/n}\left[\cos\frac{(2k\pi + \theta)}{n} + i\sin\frac{(2k\pi + \theta)}{n}\right] \quad ....(i)$$

$$k = 0, 1, 2, ..., (n - 1)$$

Equation (i) gives n distinct roots of z. These roots can be written as

$$z^{1/n} = r^{1/n}\, e^{(2k\pi + \theta)i/n};$$

$$k = 0, 1, 2, ... (n - 1) \text{ or}$$

$$z^{1/n} = r^{1/n} . e^{2k\pi i/n} . e^{\theta i/n}$$

which shows that the roots are in G.P. of first term $r^{1/n}\, e^{\theta i/n}$ and common ratio $e^{2\pi i/n}$. We exhibit the process by giving an example.

Find the fourth roots of $1 + i\sqrt{3}$

Sol. $1 + i\sqrt{3} = 2\left(\frac{1}{2} + i\sqrt{3}/2\right)$

$= 2\ [\cos \pi/3 + i \sin \pi/3]$

$= 2\ [\cos (2n\pi + \pi/3) + i \sin (2n\pi + \pi/3)]$

$\therefore \quad (1 + i\sqrt{3}\ )^{1/4} = 2^{1/4}\left[\cos\left(\frac{6n\pi + \pi}{12}\right) + i \sin\left(\frac{6n\pi + n}{12}\right)\right];$

$n = 0, 1, 2, 3.$

3

QUADRATIC EQUATIONS

Polynomial: An expression of the form $f(x) = a_0x^2 + a_1 x + a_2$ where a_0, a_1, a_2 are real constants. *i.e.*, $a_0, a_1, a_2 \in R$ and x is a variable is called a polynomial in x.

Coefficients: The constants a_0, a_1, a_2 are called coefficients of the polynomial $f(x)$.

Degree of a Polynomial: The degree of an equation involving one variable is given by the highest power of the variable in the equation after the equation has been reduced to the rational integral form.

An equation of the first degree is also called linear equation.

An equation of the second degree is also called a quadratic equation.

An equation of the third degree is also called a cubic equation.

Equation: If two different polynomials in the same variable x become equal for some values of x or a polynomial is equated to zero, then it is called an *equation*, *i.e.*, the polynomial $f(x) = a_0x^2 + a_1 x + a_2 = 0$, $a_0 \neq 0$ is an equation of degree 2.

Identity: An identity is a statement of equality between two algebraic expressions but is satisfied for all values of x. e.g., $(x-1)(x-2) \equiv x^2 - 3x + 2$ is satisfied for all values of x. The sign of identity is $\equiv$.

Quadratic Expression: $f(x) = ax^2 + bx + c$, $a \neq 0$; $a, b, c \in$ C is called a quadratic expression or function or polynomial of the second degree with complex coefficients.

If $a, b, c \in$ R, then $f(x)$ is called a *real polynomial*.

Quadratic Equation: If the polynomial $f(x)$ when equated to zero is satisfied by one or more particular values of x, then $f(x) = 0$ is called an equation.

The general form of the quadratic equation in x is $ax^2 + bx + c = 0$, when $a \neq 0$. For example $f(x) = x^2 - 3x + 2 = 0$ is a quadratic equation as it is satisfied for $x = 2$ and 1 only.

Roots of an Equation: We have, $ax^2 + bx + c = 0$

or $\quad a^2x^2 + abx + ac = 0$

or $$\left(ax+\frac{1}{2}b\right)^2 = \frac{1}{4}(b^2-4ac)$$

or $$ax+\frac{1}{2}b = \pm\frac{1}{2}\sqrt{(b^2-4ac)}$$

$\therefore$ $$x = \frac{-b\pm\sqrt{(b^2-4ac)}}{2a}$$

If α and β be the roots of the equation, and $\alpha > \beta$, then

$$\alpha = \frac{-b+\sqrt{(b^2-4ac)}}{2a}$$ and

$$\beta = \frac{-b-\sqrt{(b^2-4ac)}}{2a}$$

Obviously $$\sigma_1 = \alpha+\beta = \frac{-b}{a}$$

$$\sigma_2 = \alpha\beta = \frac{c}{a}$$

$$\alpha - \beta = \frac{\sqrt{(b^2 - 4ac)}}{a}$$

The quantity $b^2 - 4ac = \Delta$ is called the discriminant of the equation.

Type-I: In the equation of the type $ax^{2n} + bx^n + c = 0$, if x^n is put equal to y the equation becomes $ay^2 + by + c = 0$ which is a quadratic equation.

For example, solve $4^{1+x} + 4^{1-x} = 10$

Here, $\quad 4^1 \times 4^x + 4^1 \times 4^{-x} = 10$

or, $\quad 4 \times 4^x + \dfrac{4}{4^x} = 10$

or, $\quad 4 \times 4^{2x} + 4 - 10 \times 4^x = 0$

Putting $\quad 4^x = y$ we have

$$4y^2 - 10y + 4 = 0$$

or, $\quad 2y^2 - 5y + 2 = 0$

or, $\quad (y - 2)(2y - 1) = 0$

either, $\quad y = 2$ or $y = \dfrac{1}{2}$

either, $\quad 4^x = 2$ or $\dfrac{1}{2}$

either, $\quad 2^{2x} = 2^1$ or 2^{-1}

or, $\quad 2x = 1$ or -1

$$\therefore \quad x = \frac{1}{2} \text{ or } \frac{-1}{2}$$

Type-II: $az + \dfrac{b}{z} = c$, when a, b, c are constants.

For example, solve $\sqrt{\dfrac{x}{1-x}} + \sqrt{\dfrac{1-x}{x}} = \dfrac{13}{6}$

Here, $\sqrt{\dfrac{x}{1-x}}$ and $\sqrt{\dfrac{1-x}{x}}$ are reciprocal of each other.

Putting $\sqrt{\dfrac{x}{1-x}} = y$, the given equation becomes

$$y + \frac{1}{y} = \frac{13}{6} \text{ or } 6y^2 - 13y + 6 = 0$$

$$\therefore \quad y = \frac{13 \pm \sqrt{169-144}}{12} = \frac{3}{2}; \frac{2}{3}$$

But $$y = \sqrt{\frac{x}{1-x}}$$

$$\therefore \quad \sqrt{\frac{x}{1-x}} = \frac{3}{2} \text{ or } \sqrt{\frac{x}{1-x}} = \frac{2}{3}$$

On squaring, we get

$$\therefore \quad \frac{x}{1-x} = \frac{9}{4} \qquad \Big| \qquad \frac{x}{1-x} = \frac{4}{9}$$

$$\Rightarrow \quad x = \frac{9}{13} \qquad \Big| \qquad x = \frac{4}{13}$$

Hence, the solution is $\left(\frac{4}{13}, \frac{9}{13}\right)$

Type-III: Equation of the type $(x+a)(x+b)(x+c)(x+d)+k=0$ when sum of two of the quantities a, b, c, d is equal to the sum of the other two, can be solved as shown below:

$$(x+1)(x+2)(x+3)(x+4)+1=0$$

$$\Rightarrow \quad [(x+1)(x+4)][(x+2)(x+3)]+1=0 \quad [\because 1+4=2+3]$$

$$\Rightarrow \quad (x^2+5x+4)(x^2+5x+6)+1=0$$

Let $x^2+5x=y$, then

$$(y+4)(y+6)+1=0$$

$$\Rightarrow \quad y^2+10y+25=0$$

either $(y+5)^2=0$ or $y=-5$

$$\therefore \quad x^2+5x=-5$$

or $x^2+5x+5=0$

or $x = \frac{-5 \pm \sqrt{25-20}}{2} = \frac{-5 \pm \sqrt{5}}{2}$

Type-IV: Equation of the type $\sqrt{ax+b} = k$

or $\sqrt{ax+b} + \sqrt{cx+d} = k$

or $\sqrt{ax+b} + \sqrt{cx+d} = \sqrt{ex+f}$

Working Rules

(i) Square both sides.

(ii) Transpose so that the expression under radical sign is on one side.

(iii) Square both sides again and solve the resulting equation.

(iv) Test all the values of x and reject those values which do not satisfy the given equation.

For example, solve $\sqrt{2x+1} + \sqrt{3x+2}$

$$= \sqrt{5x+3}$$

Here, squaring both sides,

$$(2x+1) + (3x+2) + 2\sqrt{(2x+1)(3x+2)} = 5x+3$$

$$\Rightarrow \quad 2\sqrt{(2x+1)(3x+2)} = 0$$

$$\Rightarrow \quad \sqrt{(2x+1)(3x+2)} = 0$$

Squaring again,

$$(2x + 1)(3x + 2) = 0$$

$$\therefore \qquad x = -\frac{1}{2}, -\frac{2}{3}$$

These two values of x satisfies the above example.

Nature of the roots of the quadratic equation: The most general quadratic equation is $ax^2 + bx + c = 0$. The nature of the roots of the given quadratic equation $ax^2 + bx + c = 0$, depends upon its discriminant, D or Δ, where $\Delta = b^2 - 4ac$, $\forall\ a, b, c \in$ R.

Case-I: When $b^2 - 4ac = 0$; the roots are real, rational and equal.

Case-II: When $b^2 - 4ac < 0$; the roots are unequal and imaginary i.e., the roots are complex and conjugate to each other.

Case-III: When $\Delta = b^2 - 4ac > 0$ but not a perfect square. In this case the roots are real, irrational and unequal.

Case-IV: When $\Delta = b^2 - 4ac$ is perfect square. In this case the $\sqrt{b^2 - 4ac}$ is real and rational. The roots are real, rational and unequal.

Relation Between Roots and Co-efficients of a Quadratic Equation $ax^2 + bx + c = 0$ are

$$\alpha = \frac{-b+\sqrt{b^2-4ac}}{2a},$$

$$\beta = \frac{-b-\sqrt{b^2-4ac}}{2a}$$

Sum of the roots $= \alpha + \beta = \frac{-b}{a} = \frac{-\text{coeff. of } x}{\text{coeff. of } x^2}$

Product of the roots $= \alpha\beta = \frac{c}{a} = \frac{\text{constant term}}{\text{coeff. of } x^2}$

Symmetric function of the roots: An expression involving α and β which remains unchanged by interchanging α and β is called a symmetric function of α and β.

For example, the functions

$\alpha + \beta$, $\alpha^2 + \beta^2$, $\alpha^3 + \beta^3$, $\alpha\beta$ are symmetric function of α and β. Following results should be carefully noted:

(i) $\alpha^2 + \beta^2 = (\alpha + \beta)^2 - 2\alpha\beta$

(ii) $(\alpha - \beta)^2 = (\alpha + \beta)^2 - 4\alpha\beta$

(iii) $\alpha^2 - \beta^2 = (\alpha + \beta)(\alpha - \beta) = (\alpha + \beta)\sqrt{(\alpha + \beta)^2 - 4\alpha\beta}$

(iv) $\alpha - \beta = \left[(\alpha + \beta)^2 - 4\alpha\beta\right]^{\frac{1}{2}}$

(v) $\alpha^3 + \beta^3 = (\alpha + \beta)^3 - 3\,\alpha\beta\,(\alpha + \beta)$
$= (\alpha + \beta)\,(\alpha^2 - \alpha\beta + \beta^2)$
$= (\alpha + \beta)\,[(\alpha + \beta)^2 - 3\alpha\beta]$

(vi) $\alpha^3 - \beta^3 = (\alpha - \beta)\,(\alpha^2 + \alpha\beta + \beta^2)$

(vii) $\alpha^4 + \beta^4 = [(\alpha + \beta)^2 - 2\alpha\beta]^2 - 2\alpha^2\beta^2$

(viii) $\alpha^4 - \beta^4 = (\alpha^2 - \beta^2)\,(\alpha^2 + \beta^2)$
$= (\alpha - \beta)\,(\alpha + \beta)\,(\alpha^2 + \beta^2)$

(ix) $\alpha^5 + \beta^5 = (\alpha^3 + \beta^3)\,(\alpha^2 + \beta^2) - \alpha^2\beta^2\,(\alpha + \beta)$

Some Important Properties:

(i) If α and β are roots of $f(x) = ax^2 + bx + c = 0$, then $f(x) = a\,(x - \alpha)\,(x - \beta)$.

(ii) The equation whose roots are α, β is $x^2 - (\alpha + \beta)x + \alpha\beta = 0$.

(iii) ***One Common root:*** Two quadratic equations $f(x) = a_1x^2 + b_1\,x + c_1 = 0$ and $g(x) = a_2x^2 + b_2x + c_2 = 0$ has only one common root α if,
$a_1\alpha^2 + b_1\alpha + c_1 = 0,\ a_2\alpha^2 + b_2\alpha + c_2 = 0.$
Hence, by the method of cross-multiplication, we get, $\dfrac{\alpha^2}{b_1c_2 - b_2\,c_1} = \dfrac{\alpha}{c_1a_2 - c_2a_1} =$

$$\frac{1}{a_1b_2 - a_2b_1},\ (a_1b_2 - a_2b_1 \neq 0)$$

Thus, the required condition for one common root is $(c_1a_2 - c_2a_1)^2 = (b_1c_2 - b_2c_1)(a_1b_2 - a_2b_1)$, and if $c_1, c_2 \neq 0$ (i.e., $\alpha \neq 0$), then

we have $\alpha = \dfrac{b_1c_2 - b_2c_1}{c_1a_2 - c_2a_1}$ or $\dfrac{c_1a_2 - c_2a_1}{a_1b_2 - a_2b_1}$

(iv) ***Both Common Roots:*** Both roots of the equation $a_1x^2 + b_1x + c_1 = 0$ and $a_2x^2 + b_2x + c_2 = 0$ are common iff

$$\frac{a_1}{a_2} = \frac{b_1}{b_2} = \frac{c_1}{c_2}$$

(v) If α is a repeated root of $f(x) = 0$, then α is also a root of the equation $f'(x) = 0$.

(vi) If α is repeated common root of $f(x) = 0$ and $g(x) = 0$, then α is also a common root of the equations $f'(x) = 0$ and $g'(x) = 0$.

Polynomial Function of a Root: Let $f(x) = ax^2 + bx + c = 0$ has a root α and $g(x)$ be a polynomial in α of any degree, then $g(x)$ can be reduced to a polynomial of one degree in α.

If $q(\alpha)$ and $r(\alpha)$ be the quotient and remainder when $g(\alpha)$ is divided by $f(\alpha)$, then

$g(\alpha) = q(\alpha)\, f(\alpha) + r(\alpha),\ 0 \leq \deg r(\alpha) < 2.$

Thus, $g(\alpha) = r(\alpha)$ as $f(\alpha) = 0$

For example, Let $f(x) = 2x^2 - 3x - 1 = 0$, has a root α, then $f(\alpha) = 2\alpha^2 - 3\alpha - 1 = 0$.

Let $\quad g(\alpha) = 4\alpha^4 - 3\alpha^2 + \alpha - 1$,

then dividing $g(\alpha)$ by $f(\alpha)$, we get

$$g(\alpha) = (2\alpha^2 + 3\alpha + 4)\,(2\alpha^2 - 3\alpha - 1) + (16\,\alpha + 3)$$

Hence, $\quad g(\alpha) = 16\,\alpha + 3$ (as $2\alpha^2 - 3\alpha - 1 = 0$)

Resolution of $ax^2 + 2hxy + by^2 + 2gx + 2\,fy + c$ into two linear factors: First we consider the quadratic expression in $x, y : f(x, y)$

$$= ax^2 + 2hxy + \text{b}y^2 + 2gx + 2fy + c \qquad .....(i)$$

On equating it to zero and considering it as quadratic equation in x, we get

$$f(x, y) = ax^2 + 2x\,(hy + g) + (by^2 + 2fy + c) = 0$$

After solving this, we get

$$x = \left\{ \frac{-(hy+g) \pm \sqrt{\left[(hy+g)^2 - a\left(by^2 + 2fy + c\right)\right]}}{a} \right\}$$

or, $ax + hy + g = \pm \sqrt{[(h^2 - ab)y^2 + 2y(hg - af) + (g^2 - ac)]}$

In order that $f(x, y)$ may be the product of two linear factors of the form $lx + my + n$, the quantity under the radical sign must be perfect square and the co-efficient of y^2 i.e., $h^2 - ab > 0$.

Hence, $(hg - af)^2 = (h^2 - ab)(g^2 - ac)$ which on dividing by a, gives $abc + fgh - af^2 - bg^2 - ch^2 = 0$, and $h^2 - ab > 0$. ..... (*i*)

If $h^2 - ab = 0$, then *eqn.* (*i*) becomes perfect square.

For the necessity of the condition $h^2 - ab > 0$, consider the quadratic

$$f(x, y) = 2x^2 + 4xy + 6y^2 - 8x + 12y + 33$$

which satisfy the condition (*i*) but is not resolvable into linear factors. Here, $h^2 - ab < 0$.

(*i*) The above condition in the determinant form can be expressed as

$$\begin{vmatrix} a & h & g \\ h & b & f \\ g & f & c \end{vmatrix} = 0 \text{ and } h^2 - ab \geq 0.$$

(*ii*) This result is very useful in analytical geometry to examine whether general equation of second degree in x, y represents two real straight lines or not.

Solutions of Some quadratic and Higher Degree Inequation:

1. For $a > 0$, $x^2 < a^2$ or $|x| < a$ iff. $-a < x < a$.
2. For $a > 0$, $x^2 > a^2$ or $|x| > a$ iff. $x < -a$ or $x > a$
3. For $\alpha < \beta$, $(x - \alpha)(x - \beta) < 0$ iff $\alpha < x < \beta$.
4. For $\alpha < \beta$, $(x - \alpha)(x - \beta) > 0$, iff. $x < \alpha$ or $x > \beta$.

For example, consider $3x^2 + 2x - 1 < 0$.

The inequation is equivalent to $x^2 + \frac{3}{2}x - \frac{1}{3} < 0$.

$$\text{i.e., } \left(x + \frac{1}{3}\right)^2 - 4/9 < 0$$

By using eqn. (i), we get

$$-\frac{2}{3} < x + \frac{1}{3} < \frac{2}{3}, \text{ i.e., } -1 < x < \frac{1}{3}$$

Note: The above inequations can also be solved in the following way:

Consider $f(x) = (x - \alpha)(x - \beta), \alpha < \beta$

$f(x) = 0$ gives $x = \alpha, \beta$, then

	$x - \alpha$	$x - \beta$	$f(x)$
$x < \alpha$	$-$	$-$	$+$
$\alpha < x < \beta$	$+$	$-$	$-$
$x > \beta$	$+$	$+$	$+$

so $f(x) = (x-\alpha)(x-\beta) < 0$ for $\alpha < x < \beta$,

and $f(x) = (x-\alpha)(x-\beta) > 0$ for $x < \alpha$ or $x > \beta$

i.e., ——— × ——— × ——— ×

$f(x)$: + $\alpha - \beta$ +

The above mentioned procedure can also be used in solving higher degree inequations and rational integral inequations. The procedure involve 3 steps.

Step I: Find x by putting numerator and denominator equal to zero and convert the function into linear factors. These values of x may be termed as critical points.

Step II: Put all these points on the number line in order.

Step III: Start with + sign from extreme right and then take alternately –ve and +ve signs.

+ • − • + • − • +

Then write the answer accordingly.

For example, (1) solve $(2x+1)(x-3)(x-1) > 0$

Here, Let $f(x) = (2x+1)(x-3)(x-1)$

$f(x) = 0$ gives $x = -\frac{1}{2}, 3, 1$

Putting those roots in order, we get $x = -\frac{1}{2}, 1, 3$.

——×——×——×——

$f(x)$: $-\ \frac{1}{2}\ \ +\ 1\ \ -\ 3\ +$

So $f(x) > 0$ in $-½ < x < 1$ and $x > 3$

(2) Solve $f(x) = \dfrac{(x-1)(x+2)}{(x-5)} \le 0$

Here critical points are 1, –2, 5.

$f(x)$: – –2 + 1 – 5 +

and $f(x) = 0$ for $x = 1$ and -2, $x \neq 5$ as $(x - 5)$ is in denominator.

So, $f(x) \le 0$ for $x \le -2$ and $1 \le x < 5$

This can also be expressed as solution set

$$x \in [-\infty, -2] \cup (1, 5)$$

Sign of Quadratic Expression $f(x)$: $ax^2 + bx + c$.

Consider $f(x) = ax^2 + bx + c$, $a \neq 0$, $a, b, c \in R$

Case I: $\Delta = b^2 - 4ac < 0$, i.e., $f(x) = 0$ has imaginary roots.

$$f(x) = a\left\{x^2 + \frac{b}{a}x + \frac{c}{a}\right\}$$

$$= a\left\{\left(x + \frac{b}{2a}\right)^2 + \frac{4ac - b^2}{4a^2}\right\}$$

as $\Delta < 0$, $\dfrac{4ac - b^2}{4a^2} > 0$

$$\text{so } \left\{\left(x + \frac{b}{2a}\right)^2 + \frac{4ac - b^2}{4a^2}\right\} > 0.$$

Thus $f(x) > 0$ for all $x \in R$ if $a > 0$, $f(x) < 0$ for all $x \in R$ if $a < 0$, *i.e.*, $f(x)$ and a have same sign for all $x \in R$.

Case II: $\Delta = 0$ then $f(x) = a\,(x + b/2a)^2$.

As $(x + b/2a)^2 \geq 0$ for all $x \in R$.

$\therefore$ We have $f(x) \geq 0$ for all $x \in R$ if $a > 0$.

$f(x) < 0$ for all $x \in R$ if $a < 0$.

Note: From case I and II, it is clear that
(i) $f(x) \geq 0$ for all $x \in R$ if $\Delta \leq 0$ and $a > 0$ and
(ii) $f(x) < 0$ for all real $x \in R$ if $\Delta \leq 0$; and $a < 0$.

For example, find a if $2x^2 + ax + 5 \geq 0$ for all real x. Here, from the above discussion, it is clear that $\Delta \leq 0$ as co-efficient of $x^2 = 2 > 0$.

$$\therefore\ a^2 - 40 \leq 0,\ a^2 \leq \left(2\sqrt{10}\right)^2$$

$$\therefore\ -2\sqrt{10} \leq a \leq 2\sqrt{10}$$

Case III: $\Delta > 0$. In this case roots are real and distinct. Let α, β be the roots and $\alpha < \beta$. Then $f(x) = a\,(x - \alpha)\,(x - \beta)$.

If $a > 0$, then

——×——×——

$f(x)$: > 0; $\alpha < 0, \beta > 0$

i.e., if $a < x < \beta$; $f(x) < 0$,

if $x < \alpha$ or $x > \beta$; $f(x) > 0$.

and if $x < 0$, then

—×—×—

$f(x) : < 0;\ \alpha > 0,\ \beta < 0$

Hence, from the above discussion, it is clear that a and $f(x)$ have the same sign except when roots are real and x is lying between them.

Location of roots (Interval in which roots lie)

Let $f(x) = ax^2 + bx + c = 0$, $a, b, c \in \mathrm{R}$, $a > 0$ and α, β be the roots.

(*i*) Both roots of $f(x) = 0$ are greater than some fixed number k.

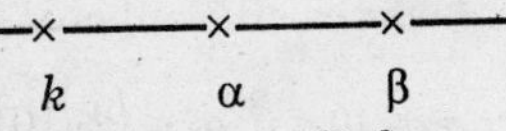

Roots must be real and as k does not lie between α and β. So $\Delta \geq 0$, $f(x) > 0$, also $\alpha + \beta > 2k$, *i.e.*, $\Delta \geq 0$, $f(x) > 0$, $-(b/a) > 2k$.

It should be noted that, both roots are less than K, if $\Delta \geq 0$, $f(k) > 0$, $-(b/a) < 2k$

(*ii*) One root is less than k and other is greater than k.

—×—×—×—

α $\quad k \quad$ β

Here roots must be real and distinct so $\Delta > 0$, $f(k) < 0$.

(*iii*) Exactly one root lies in the interval (k_1, k_2) *i.e.*,

$$\beta \quad k_1 \quad \alpha \quad k_2 \quad \beta$$

$$\beta < k_1 < \alpha < k_2 \text{ or } k_1 < \alpha < k_2 < \beta,$$

then, one of $f(k_1)$ or $f(k_2)$ is negative and other is positive so $\Delta > 0$, $f(k_1)\, f(k_2) < 0$.

(*iv*) If both roots are confined between the numbers p, q,

$$p \quad \alpha \quad \beta \quad q$$

Here, roots must be real so $\Delta \geq 0$, $f(p) > 0$, $f(q) > 0$,

$$p < \frac{\alpha+\beta}{2} < q, \text{ i.e., } p < -\frac{b}{2a} < q$$

(*v*) If $f(p)$ and $f(q)$ have opposite sign, $f(x) = 0$ must have at least one root between p and q *i.e.*, there exists at least one α such that $p < \alpha < q$, $f(x) = 0$.

This result is also applicable for any polynomial equation of any degree and is due to *Rolle*.

(*vi*) If $f(x)$ be any polynomial such that $f(a) = f(b) = 0$, then $f'(x) = 0$ has at least one root between a and b *i.e.*, there exists at least one c such that

$$a < c < b \text{ and } f'(c) = 0$$

This result is a particular version for polynomial equations of a more general theorem calld Rolle's theorem.

Equation of higher degree: The equation $f(x) = a_0x^n + a_1 x^{n-1} + a_2 x^{n-2} + ...+ a_{n-1} x + a_n = 0$. $a_0, a_1 ..., a_n \in C$ (or R), $a_0 \neq 0$ is a polynomial equation of degree n. It has n and only n roots. Let $\alpha_1, \alpha_2 ....., \alpha_n$ be n roots, then

$$\sigma_1 = \Sigma\, \alpha_1 = -\frac{a_1}{a_0},\ \sigma_2 = \Sigma\, \alpha_1\, \alpha_2 = \frac{a_2}{a_0} \text{ etc.}$$

In general $\sigma_r = \Sigma\alpha_1 ..... \alpha_r = (-1)^r \dfrac{a_r}{a_0}$

In particular, if α, β, γ be roots of

$$ax^3 + bx^2 + cx + d = 0,$$

We have

$$\sigma_1 = \alpha + \beta + \gamma = -b/a$$
$$\sigma_2 = \alpha\beta + \beta\gamma + \gamma\alpha = c/a$$
$$\sigma_3 = \alpha\beta\gamma = (-d/a).$$

4

SEQUENCE AND SERIES

Sequence: A sequence is a special class of function whose domain is the set of N of all natural numbers and range is any set, i.e., A function $f : N \rightarrow S$ is called a sequence. If the range of a sequence is any subset of real numbers, then the sequence is called a real sequence.

Therefore, a real sequence is a function from the set N of natural numbers to the set $S \subset R$ of real numbers.

Different Ways of Describing a Sequence

(*i*) A sequence may be described by listing its first few elements, till we get a rule for writing down the other different members of the sequence, e.g. $\left\langle 1, \frac{1}{2}, \frac{1}{3}, \frac{1}{4}, \ldots\ldots \right\rangle$ is the sequence whose n^{th} term is $\frac{1}{n}$.

(*ii*) Another way of representing the member of the sequence is to specify the rule for its nth term, e.g., the sequence $\left\langle 1, \frac{1}{2}, \frac{1}{3}, \ldots\ldots \right\rangle$ can be rewritten as $\langle f_n \rangle$ where $f_n = \frac{1}{n}$ for all $n \in N$.

Here $f_n = \frac{1}{n}$, gives the rule for the nth term of the sequence.

(*iii*) Lastly, a sequence can also be described by specifying the first term and storing a rule for determining f_n for all $n \geq 1$ in terms of the term f_1, f_2, f_3 ..... e.g., sequence $\langle f_n \rangle$ for which $f_0 = 1, f_1 = 2$ and $f_n = \frac{1}{2}(f_{n-1} + f_{n-2})$ for all $n \geq 2$ i.e., $\left\langle 1, 1, \frac{8}{2}, \frac{7}{4} \ldots\ldots \right\rangle$

Constant Sequence: The sequence $\langle f_n \rangle$ where $f_n = C \in R$, for all $n \in N$ is called constant sequence. In the case $\langle f_n \rangle = \langle c_1\ c_1\ c_1 \ldots\ldots \rangle$

$\therefore$ Range of $f = \{c\}$ = singleton c = a finite set

It should be noted that

(*i*) It is not necessary that all the terms of the sequence should be distinct.

(*ii*) Care must be taken in distinguishing the range of the sequence and the sequence itself. e.g., the sequence (f_n), where $f_n = (-1)^{n-1}$ for all $n \in \mathrm{N}$ is given by

$$\langle f_n \rangle = \langle 1, -1, 1, -1, 1, -1..... \rangle$$

Here, the range of $f = \{1, -1\}$

(*iii*) A sequence by definition is always an infinite set, while the range of the sequence may be finite or infinite, e.g., the sequence $\langle f_n \rangle$ for which $f_n = 1$ for all $n \in \mathrm{N}$.

Thus, $\langle f_n \rangle = \langle 1, 1, 1, ... \rangle$, where as range of $f = \{1\}$ which is a finite set.

(*iv*) A sequence can also be defined as a succession of terms arranged in a definite order and formed according to a definite law, i.e., the set $\langle f_1, f_2, ... f_n ... \rangle$ is called a sequence, where $f_1, f_2, ... f_n ...$ are real numbers arranged in a definite order and formed according to some law.

For example, the set $\langle 1, 4, 9, 16, 25, \ldots \rangle$ is sequence and the law here is

$f_n = n^2$ for all $n \in \mathrm{N}$

Series: A series can be defined as the succession of terms formed and arranged according to some definite law of rule. Let $\langle u_n \rangle$ be a real sequence. The expression of the form $u_1 + u_2 + \ldots + u_n \ldots$ is called a series and is symbolically expressed as

$\sum_{n=1}^{\infty} u_n$ or simply $\sum u_n$. Then real number $u_1, u_2 \ldots$ are known as first, second,... respectively of the series $\sum u_n$.

It should be noted that,
if $\{t_1, t_2, t_3, t_4, \ldots\}$ is a sequence, then
$S = t_1 + t_2 + t_3 + \ldots$ is called the corresponding series.
A sequence is said to be an Arithmetic sequence or Arithmetic progression, if the difference of each term after the first term and the preceding term is constant.

Arithmetic Progression: An arithmetic progression (A.P) is a sequence whose terms increase or decrease by a fixed number. Thus i sequence $\{t_1, t_2, t_3...\}$ is such that $t_n - t_{n-1}$ is a constant for all $n \in$ N, it is an A.P. This fixed number is called the common difference of the A.P. For example, 1, 3, 5, 7, 9, ... is an A.P. with common difference = 2.

$$[\because 3 - 1 = 5 - 3 = 7 - 5 = 9 - 7 = ... = 2]$$

The standard A.P. is a, $(a + d)$, $(a + 2d)$, $(a + 3d)$... where a is the first term and d is the common difference.

***n*th term of an A.P.:** If a is the first term and d the common difference of an A.P., then its nth term t_n is given by $t_n = a + (n - 1)d$.

Sum of *n* terms of an A.P.: If a is the first term and d the common difference of an A.P., then sum of n terms (denoted by S_n) is given by

$$S_n = \frac{n}{2}[2a + (n - 1)d] = \frac{n}{2}(a + l)\text{, where } l \text{ is the}$$

last term.

Arithmetic Mean: Arithmetic mean (A.M.) of any two numbers a and b is given by $(a + b)/2$, i.e., the arithmetic mean between two given numbers is equal to half of their sum. If $x_1, x_2, \ldots x_n$ are n numbers, then their A.M. is given by $(x_1 + x_2 + \ldots + x_n)/n$

***n* arithmetic means:** The numbers $A_1, A_2, \ldots A_n$ are said to be n arithmetic means between a and b if

$$a, A_1, A_2, A_3, \ldots A_n, b \text{ are an A.P.}$$

Hence, $b = t_{n+2} = a + (n + 1)\,d.$

$$\therefore \quad d = \frac{b-a}{n+1} \text{ and hence, } A_1 = a + \frac{b-a}{n+1}, \ldots$$

$$A_r = a + \frac{r(b-a)}{n+1}$$

Important Deductions:

(i) If a fixed number is added or subtracted to each term of a given A.P., then the resulting series is also an A.P. and its common difference remains the same.

(ii) If each term of an A.P. is multiplied by a fixed constant or divided by a fixed non-zero constant, then the resulting series is also an A.P.

For example, an A.P. is multiplied by the same number, i.e., ak, $(a + d)\,k$, $(a + 2d)\,k \ldots$ then the resulting sequence is also an A.P.

(iii) If $x_1 + x_2 + x_3 + \ldots$ and $y_1 + y_2 + y_3 + \ldots$ are two A.P., then $x_1 \pm y_1, x_2 \pm y_2, x_3 \pm y_3 + \ldots$ are also in A.P.

(iv) If the sum of three numbers in A.P. is given; take them as $a - d, a, a + d$.

(v) If the sum of four numbers in A.P. is given; take them as $a - 3d, a - d, a + d, a + 3d$.

(vi) If the sum of the five numbers in A.P. is given; take them as $a - 2d, a - d, a, a + d, a + 2d$.

Geometric Progression (G.P.): A sequence is said to be a G.P. or G.S. if the ratio of any term, after the first term, to its proceeding term is constant. This constant ratio is called the common ratio of the G.P. and is usually denoted by r.

It follows from the definition that no term of a G.P. can be zero.

The standard G.P. is

$$a + ar + ar^2 + ar^3 + \ldots \left[\because \frac{ar}{a} = \frac{ar^2}{ar} = \ldots = r\right]$$

where a is the first term of a G.P. and r is the common ratio of G.P.

Note: It should be noted that

(a) for $r = 1$, the G.P. is $a + a + a + \ldots + a$ (n times) and its sum is $S_n = na$ (If $r = 1$);

(b) If we multiply any term of a G.P. by a common ratio, we get the next following term and if we divide any term of the G.P. by the common ratio, we get the proceeding term.

***n*th term of a G.P.:** If a is the first term and r is common ratio of a G.P., then n^{th} term

$$= t_n = ar^{n-1}$$

Sum of *n* terms of a G.P.: If a is the first term and r is common ratio of a G.P., then sum of n terms, denoted by S_n is given by

$$S_n = \frac{a(r^n - 1)}{r - 1} = \frac{a(1 - r^n)}{1 - r} \text{ if } r < 1$$

Sum of Infinite G.P.: If $-1 < r < 1$ i.e., $|r| < 1$, then the sum of the infinite G.P.

$$a + ar + ar^2 + ... = a/(1 - r).$$

Geometric mean (G.M.): If $a, b > 0$, then the geometric mean of a and $b = \sqrt{ab}$.

If $a_1, a_2 ..., a_n > 0$, then their geometric mean is given by $(a_1, a_2 ... a_n)^{1/n}$.

If n geometric mean $g_1, g_2, ... g_n$ are to be inserted between two positive real numbers a and b then $a, g_1, g_2, ..., g_n, b$ are in G.P.

so, $\quad b = ar^{n+1}$ i.e., $r = (b/a)^{1/(n+1)}$ and

then $\quad g_1 = ar, g_2 = ar^2, ..., g_n = ar^n$

Some Important Deductions:

(i) If each term of a G.P. is multiplied or divided by some fixed non-zero number, then resulting sequence is also a G.P.

(ii) If $x_2, x_2, x_3 \ldots$ and $y_1, y_2, y_3, \ldots$ are two G.P. then

$x_1y_1, x_2y_2, x_3y_3, \ldots$ and $\frac{x_1}{y_1}, \frac{x_2}{y_2}, \frac{x_3}{y_3}, \ldots$ are

also in G.P.

(iii) If $x_1, x_2, x_3 \ldots$ is a G.P. of positive terms, then $\log x_1, \log x_2, \log x_3, \ldots$ is an A.P. and vice versa.

(iv) Three terms of a G.P. can be taken as $a/r, a, ar$ and four terms in G.P. as $a/r^3, a/r, ar, ar^3$. This presentation is useful if product of terms is involved in the problem. In other problems, terms should be taken as $a, ar, ar^2 \ldots$.

Harmonic Progression (H.P.): The sequence $x_1, x_2, x_3 \ldots, x_n \ldots$, where $x_n \neq 0$ for each n, is said to be in H.P. if the sequence $1/x_1, 1/x_2, 1/x_3 \ldots$ is an A.P. It should be noted that no term of H.P. can be zero.

***n*th term of a H.P.:** If $x_1, x_2, x_3, \ldots$ is a H.P., then

the corresponding A.P. is $\frac{1}{x_1}, \frac{1}{x_2} \ldots$.

Hence, $a = \frac{1}{x_1}$ and $d = \frac{1}{x_2} - \frac{1}{x_1}$, then nth term of H.P. $(x_n) = \frac{1}{a+(n-1)d}$

Harmonic Mean (H.M.): If a and b are two non-zero numbers, then the harmonic mean of a and b, denoted by H is given by $\frac{1}{H} = \frac{1}{2}\left(\frac{1}{a}+\frac{1}{b}\right)$ or $H = \frac{2ab}{a+b}$. If $a_1, a_2, a_3, ..., a_n$ be n non-zero numbers, then their harmonic mean H is given by

$$\frac{1}{H} = \frac{1}{2}\left(\frac{1}{a_1}+\frac{1}{a_2}+...+\frac{1}{a_n}\right)$$

If $H_1, H_2, ... H_n$ are n harmonic means between two non-zero numbers a and b, then

$\frac{1}{a}, \frac{1}{H_1}, \frac{1}{H_2}, \frac{1}{H_3}, \frac{1}{H_4}, ... \frac{1}{H_n}, \frac{1}{b}$ are in A.P.

Arithmetico-Geometric Series (A.G.S.): A series in which each term is obtained by multiplying the corresponding terms of an arithmetic and geometric series is defined as an Arithmetico-Geometric series.

Let $a + (a + d) + (a + 2d) + ... + [a + (n - 1) d]$ is a standard A.P.

and $1 + r + r^2 + ... + r^{n-1} + ...$ is a standard G.P.

Then Arithmetico-Geometric series

$$= a + (a + d) r + (a + 2d) r^2 + ... + \left(a + \overline{n-1}d\right) r^{n-1} + ...$$

***n*th term of A.G.S.:** The nth term of A.G.S. is $\{a + (n - 1) d\} r^{n-1}$

Sum of n terms of A.G.S.: The sum of n terms of this series

$$S_n = \frac{a}{1-r} + \frac{dr\left(1-r^{n-1}\right)}{(1-r)^2} - \frac{[a+(n-1)d]r^n}{1-r}$$

Sum to infinity $S = \dfrac{a}{1-r} - \dfrac{dr}{(1-r)^2}$

Formulae for Σn, Σn^2, Σn^3

(i) $\Sigma n = \sum_{r=1}^{n} r = 1 + 2 + 3 + ... + n = \frac{1}{2} n (n + 1)$

(ii) $\Sigma n^2 = \sum_{r=1}^{n} r^2 = \frac{1}{6} n(n+1)(2n+1)$

(iii) $\Sigma n^3 = \sum_{r=1}^{n} r^3 = \left[\frac{1}{2} n(n+1)\right]^2$

Connection between A.M., G.M. and H.M.: If A, G and H are respectively the A.M., G.M., and H.M. of two numbers a and b then

(i) $AH = G^2$ (ii) $A \geq G \geq H$

5

PERMUTATION AND COMBINATION

1. **Factorial Notation:** The continued product of the first n natural numbers is defined by the symbol $\underline{|n}$ or $n!$ and is read as 'Factorial n'.

   $n! = 1.2.3. \ldots n$

   $n! = n\,(n-1)! = n\,(n-1)\,(n-2)!$

2. **The Sum Rule:** Suppose a work A can occur in m ways and B can occur in n ways and both cannot occur simultaneously. Then A or B (at least one of them) can occur in $(m + n)$ ways. This rule is also applicable for two or more exclusive events.

3. **The Product Rule:** Suppose there are two works A and B. Let A can occur in m ways and B in n ways. Suppose that the ways for A and B are not related in the sense that B

occur in n ways regardless the outcome of A, then both A and B occur in mn ways.

For example, let there are two questions A and B which can be solved by two methods and 3 methods respectively. Then, A or B can be solved in $2 + 3 = 5$ ways and boh A and B in $2 \times 3 = 6$ ways.

4. **Permutation:** Each of the arrangements that can be made by taking some or all of a number of dissimilar things (objects) is called *Permutation*.
5. **Combination:** Each of the different group or selection which can be made by taking some or all of a number of things (irrespective of order) is called a *combination*.

   Derangements: Any change in the given order of the things is called a derangements.
6. The number of permutations of n distinct things taken all at a time $= n!$
7. The number of permutations of n dissimilar things taken r at a time when each thing can be repeated any number of times $= n^r$.
8. Number of permutations of n distinct things taken r at a time, $0 \leq r \leq n$

$$= n(n-1)(n-2) \ldots (n-r+1)$$
$$= n!/(n-r)! = {}^nP_r.$$

This is equivalent to filling r places by r objects taking from n distinct objects.

9. The number of permutations of n distinct things taken r at a time when p particulars things always occur $= {}^{(n-p)}C_{(r-p)} \cdot r!$.

10. Number of permutations (arrangements) of n distinct things taken r at a time when p particular things never occur

$$= {}^{(n-p)}C_r \cdot r!$$

11. Number of permutations of n things, taken all at a time when p_1 are alike of one kind, p_2 are alike of second kind, ... p_r of them are alike of the rth kind $p_1 + p_2 + ... + p_r \leq n$, and remaining things are all different $= n!/\{p_1!\, p_2! \dots p_r!\}$.

12. The number of combinations of n objects taken r at a time, $0 \leq r \leq n$.

$$= {}^nC_r = \frac{n!}{(n-r)!r!}$$

13. The number of combinations of n distinct objects taken r at a time when any object may be repeated any number of times.

$$= \text{coefficent of } x^r \text{ in } (1 + x + x^2 + x^3 + ... + x^r)^n$$

$$= \text{coefficient of } x^r \text{ in } (1 - x)^{-n} = {}^{n+r-1}C_r$$

14. Number of combinations of n distinct things taken r at a time when p particular things always occur

$$= {}^{(n-p)}C_{(r-p)}$$

15. Number of combinations of n distinct things taken r at a time when p particular things never occur $= {}^{(n-p)}C_r$.

16. The number of combinations if some or all n things be taken at a time

$$= {}^nC_1 + {}^nC_2 + ... + {}^nC_n = 2^n - 1.$$

17. If there are p_1 objects of one kind, p_2 objects of second kind, ..., p_n objects of nth kind, then the number of ways of choosing r objects out of these $(p_1 + p_2 + ... + p_n)$ objects

$$= \text{coefficient of } x^r \text{ in } \left(1 + x + ... + x^{p_1}\right)$$

$$\left(1 + x + ... + x^{p_2}\right) ... \left(1 + x + ... x^{p_n}\right)$$

If one object of each kind is to be included in such a collection then the number of ways of choosing r objects.

$$= \text{coefficient of } x^r \text{ in the product}$$

$$\left(x + x^2 + ... + x^{p_1}\right)\left(x + x^2 + ... + x^{p_2}\right) ...$$

$$\left(x + x^2 + ... + x^{p_n}\right).$$

18. Total number of ways to make a selection by taking some or all of $p_1 + p_2 + \ldots p_r$ things where p_1 are alike of one kind, p_2 are alike of second kind, ..., p_r are alike of rth kind is

$$= (p_1 + 1)(p_2 + 1)\ldots(p_r + 1) - 1$$

19. The number of ways in which n distinct objects can be split into three groups containing respectively r, s and t objects, r, s and t are distinct and $r + s + t = n$, is given by

$$^{n}C_r \, ^{n-r}C_s \, ^{n-r-s}C_t = \frac{n!}{r!.s!.t!}$$

20. If 3n things are to be divided into the three equal groups, then the number of ways

$$= \frac{(3n)!}{n!.n!.n!.3!}$$

21. If $3n$ things are to be divided equally between 3 persons (*i.e.*, division of $3n$ things into 3 equal groups with permutation of groups),

then the number of ways $= \dfrac{(3n)!}{(n!)^3}$

22. If n things form an arrangement in a row, then the number of ways in which they can be deranged so that no one of them occupies its original place is

$$n!\left(1-\frac{1}{1!}+\frac{1}{2!}-\frac{1}{3!}+...+(-1)^n.\frac{1}{n!}\right)$$

23. The number of circular permutations of n different things taken all at a time = $(n-1)!$

24. The number of arrangements of n persons on a round table = $(n-1)!$

25. The number of arrangement of n flowers to make a garland = $\frac{1}{2}$ $(n-1)!$

26. ${}^nC_r = {}^nC_{n-r}$

27. ${}^nC_r + {}^nC_{r+1} = {}^{n+1}C_{r+1}$

28. ${}^nC_r = {}^nC_s \Rightarrow r = s$ or $r + s = n$.

29. Greatest Value of ${}^nC_r$

${}^nC_r$ is greatest when $r = n/2$ if n is even.

$r = (n-1)/2$ or $(n+1)/2$ if n is odd.

30. Distinction between problems on "permutation and combination"

(*a*) In problems on arrangements, ordering, permutations, arranging, line up, we use the formula for Permutations.

(b) In problems on selection, subset, committee, groups, choice, we use the formula for Combinations.

(c) When to multiply and when to add, Multiply when there is 'AND' in the problem and 'Add' when there is 'OR'.

6

BINOMIAL THEOREM

Binomial Expression: A binomial expression is an algebraic expression consisting of two terms connected by plus (+) or minus (–) sign. For example, $2x + 3y$, $3x + 9y$, $2x - 5y$ are binomial expressions in x and y.

The Factorial Function: For $n \in N$, factorial of n, denoted by $n!$ is defined by

$$n! = n.(n-1)(n-2)\ldots 3.2.1.$$

It is also assumed that $0! = 1$

Remarks:

(a) $n! = n.(n-1)! = n(n-1)(n-2)!$ etc.

(b) $1/(-n)! = 0$

Binomial Theorem: For any positive Integer n

$(x+a)^n = {}^nC_0 x^n + {}^nC_1 x^{n-1} a + {}^nC_2 x^{n-2} a^2 + \ldots + {}^nC_r x^{n-r}. a^r + \ldots + {}^nC_n a^n$.

where the constant ${}^nC_0$, ${}^nC_1$, ${}^nC_2$, ... ${}^nC_n$ are binomial coefficient and ${}^nC_r = \dfrac{n!}{r!(n-r)!}$

(a) ***Expansion of (x − a)ⁿ:*** If we put $a = -a$ in the above theorem, we get

$$(x-a)^n = {}^nC_0\, x^n - {}^nC_1\, x^{n-1} a + {}^nC_2\, x^{n-2} a^2 + \ldots + (-1)^n\, {}^nC_n\, a^n.$$

(b) ***Expansion of (1 + x)ⁿ:*** If we put $x = 1$ and $a = x$ then

$$(1+x)^n = {}^nC_0 + {}^nC_1 x + {}^nC_2\, x^2 + \ldots + {}^nC_r\, x^r + \ldots + {}^nC_n\, x^n$$

$$= 1 + nx + \frac{n(n-1)}{2!}\, x^2 + \ldots +$$

$$\frac{n(n-1)(n-2)\ldots(n-r+1)}{r!}\, x^r + \ldots + x^n.$$

(c) ***Expansion of (1 − x)ⁿ:*** If we put $x = 1$ and $a = -x$ in the above theorem.

$$(1-x)^n = 1 - {}^nC_1\, x + {}^nC_2 x^2 - \ldots + (-1)^r\, {}^nC_r\, x^r + \ldots + (-1)^n\, x^n.$$

Important Points: For the binomial exprssion of $(x + y)^n$, where n is a positive integer.

(i) The expansion on the right hand side contains $(n + 1)$ terms which is one more than the index of the Binomial.

(ii) In each term of the expansion the sum of the exponents (power of x and y) is n (i.e., each term is of degree n).

(iii) The binomial coefficients of terms from the beginning and the end are equal since ${}^nC_r = {}^nC_{n-r}$

i.e., since ${}^nC_r = {}^nC_{n-r}$

$\therefore \quad {}^nC_0 = {}^nC_n, {}^nC_1 = {}^nC_{n-1}, {}^nC_2 = {}^nC_{n-2}$ ... etc.

Hence, the coefficients from the beginning and end are equal.

General term in the expansion of $(x + y)^n$: In the Binomial expansion of $(x + y)^n$, the $(r + 1)$th term is called the general term, and denoted by T_{r+1}. Thus $T_{r+1} = {}^nC_r x^{n-r} y^r$.

For example, the 5th term from the end in

the expansion of $\left(\frac{x^3}{2} - \frac{2}{x^2}\right)^9$ is $T_{n-p+2} = T_{9-5+2}$

$= T_C \; [\because n = 9, p = 5]$

Middle term of terms in the expansion of $(x + y)^n$: (Case I): If n is even, then there will be only one middle term in the expansion, which is $(n/2 + 1)$th term.

$\therefore$ The middle term $= (n/2 + 1)$th term

$= {}^nC_{n/2} \, x^{n/2} \, y^{n/2}$.

For example, if $n = 8$, then the middle term is $(8/2 + 1) = 5$th term *i.e.*, T_5.

(Case II): If n is odd, then there will be two middle terms in the expansion which are $\frac{1}{2}(n+1)$ and $\frac{1}{2}(n+3)$th terms.

i.e., Middle terms are

$$T_{(n+1)/2} = {}^nC_{(n-1)/2} . x^{(n+1)/2} . y^{(n-1)/2}$$
$$T_{(n+3)/2} = {}^nC_{(n+1)/2} . x^{(n-1)/2} . y^{(n+1)/2}$$

For example, if $n = 7$, then the number of terms in the expansion of $(x + y)^7$ is $7 + 1 = 8$

$\therefore$ Middle terms are 4th and 5th terms

(i.e. T_4 & T_5)

To find $(m + 1)$th term from the end: In the Binomial expansion of $(x + y)^n$, the $(m + 1)$th term from the end $= (n - m + 1)$th term from the beginning $= T_{n-m+1}$

Coefficient of a particular power of x in $(1 + x)^n$

Working Rule: *Step I:* Write down the general term T_{r+1} and simplify.

Step II: Equate the index of x in T_{r+1} to the given power and find r.

Step III: Substitute the value of r in the general term and get the term and its coefficient.

Illustration: Find the coefficients of x^{32} in the expansion of $\left(x^4 - \frac{1}{x^3}\right)^{15}$

Sol.: The general term in the equation is

$$T_{r+1} = {}^{15}C_r\,(x^4)^{15-r}\left[-\frac{1}{x^3}\right]^r$$

$$= (-1)^r\,{}^{15}C_r\,\frac{x^{60-4r}}{x^{3r}}$$

$$= (-1)^r\,{}^{15}C_r\,x^{60-7r}$$

It will involve x^{32} if $60 - 7r = 32 \quad \therefore r = 4$

$\therefore x^{32}$ occur in the 5th term

$$\text{Coefficients of } x^{32} = (-1)^4\,{}^{15}C_4 = \frac{15\times14\times13\times12}{4\times3\times2\times1}$$

$$= 1365$$

Term independent of x in $(1 + x)^n$

Working Rule: *Step I:* Write down the general term T_{r+1} and simplify.

Step II: Equate the index of x into zero and solve for r.

Step III: Substitute this value of r in the general term T_{r+1} to get term independent of x.

Illustration: Find the term independent of x in the expansion of $\left(x^2+\frac{1}{x}\right)^{12}$.

Sol.: Let T_{r+1} be the term independent of x in $\left(x^2+\frac{1}{x}\right)^{12}$

$$\therefore \quad T_{r+1} = {}^{12}C_r\,(x^2)^{12-r}\left(\frac{1}{x}\right)^r$$

$$= {}^{12}C_r\,x^{24-2r}\,\frac{1}{x^r}$$

$$\Rightarrow \quad T_{r+1} = {}^{12}C_r\,x^{24-3r} \qquad \text{... (i)}$$

Now x is to have power zero.

$$\therefore \quad 24-3r=0 \Rightarrow 3r=24 \Rightarrow r=8$$

Putting $r=8$ in eqn. (i) we get

$$T_{8+1} = {}^{12}C_8\,x^0 = \frac{12!}{8!\;4!}$$

$$= \frac{12\times11\times10\times9}{4\times3\times2}$$

$$= 495$$

Number of terms in the expansion of $(x + y + z)^n$, where n is a positive integer, is $\frac{1}{2}(n + 1)(n + 2)$

Properties of Binomial Coefficients:

(*i*) $C_0 + C_1 + C_2 + ... + C_n = 2^n$

(*ii*) $C_0 + C_2 + C_4 + ... = C_1 + C_3 + C_5 + ... = 2^{n-1}$

(*iii*) $C_1 + 2C_2 + 3C_3 + ... + {}^nC_n = n.\ 2^{n-1}$

(*iv*) $C_1 - 2C_2 + 2C_3 - ... = 0.$

(*v*) $C_0 + 2C_1 + 3C_2 + ... + (n + 1)\ C_n = (n + 1)\ 2^{n-1}$

(*vi*) $C_0C_r + C_1\ C_{r-1} + ...$

$+ C_{n-r}\ C_n = (2n)!/\{(n - r)!.(n + r)!\}$

(*vii*) $C_0^2 + C_1^2 + C_2^2 + ... + C_n^2 = (2n)!/(n!)^2$

(*viii*) $C_0^2 - C_1^2 + C_2^2 - C_3^2 ... =$

$$\begin{cases} 0, & \text{if } n \text{ is odd} \\ (-1)^{n/2}\ {}^nC_{n/2} & \text{if } n \text{ is even} \end{cases}$$

Binomial Theorem (for any index): For any rational index n $(n \neq 0)$

$$(1+x)^n = 1 + nx + \frac{n(n-1)}{2!}x^2 + \frac{n(n-1)(n-2)}{3!}x^3 + \ldots + \frac{n(n-1)(n-2)\ldots(n-r+1)}{r!}x^r + \ldots$$

Remember

(*i*) If n is a negative integer or a rational fraction, then the number of terms in the above expansion is infinite.

(*ii*) If n is a negative integer or a rational fraction, then this expansion is valid for $|x| < 1$ i.e., for $-1 < x < 1$.

(*iii*) If n is positive integer, then the number of terms in the expansion is $(n + 1)$ *i.e.,* finite, and

$$T_{r+1} = \frac{n(n-1)(n-2)\ldots(n-r+1)}{r!}x^r$$

(*iv*) If n is a positive integer, then $(1+x)^{-n}$

$$= 1 - nx + \frac{n(n+1)}{2!}x^2 - \frac{n(n+1)(n+2)}{3!}x^3 + \ldots$$

$$= 1 + (-1)^n\, {}^nC_1\, x + (-1)^2\, {}^{n+1}C_2\, x^2 + (-1)^3\, {}^{n+2}C_3\, x^2 + \ldots + (-1)^r\, {}^{n+r-1}C_r\, x^r + \ldots$$

(*v*) If n is a positive integer, then

$$(1-x)^{-n} = 1 + nx + \frac{n(n+1)}{2!}x^2 + \frac{n(n+1)(n+2)}{3!}x^3 + \ldots$$

$$= 1 + {}^{n}C_1\, x + {}^{n+1}C_2\, x^2 + {}^{n+2}C_3\, x^3 + \ldots + {}^{n+r-1}C_r\, x^r + \ldots$$

General term

(*i*) In the expansion of $(1 + x)^n$, the general term *i.e.*, the $(r + 1)$th term

$$T_{r+1} = \frac{n(n-1)(n-2)\ldots(n-r+1)}{r!}x^r$$

(*ii*) If n is a positive integer, then in the expansion of $(1 + x)^{-n}$

$$T_{r+1} = (-1)^r\, {}^{n+r-1}C_r\, x^r$$

(*iii*) If n is positive integer, then in the expansion of $(1 - x)^{-n}$

$$T_{r+1} = {}^{n+r-1}C_r\, x^r$$

Important Deductions:

(*i*) $(1-x)^{-1} = 1 + x + x^2 + x^3 + \ldots + x^r + \ldots$

(*ii*) $(1+x)^{-1} = 1 - x + x^2 - x^3 + \ldots + (-1)^r x^r + \ldots$

(*iii*) $(1-x)^{-2} = 1 + 2x + 3x^2 + 4x^3 + ... + (r+1)\, x^r + ...$

(*iv*) $(1+x)^{-2} = 1 - 2x + 3x^2 - 4x^3 + ... + (-1)^r (r+1)x^r + ...$

(*v*) $(1-x)^{-3} = 1 + 3x + 6x^2 + 10x^2 + ... + \dfrac{(r+1)(r+2)}{2} x^r + ...$

(*vi*) $(1+x)^{-3} = 1 - 3x + 6x^2 - 10x^3 + ... + (-1)^r \dfrac{(r+1)(r+2)}{2} x^r + ...$

7

EXPONENTIAL AND LOGARITHMIC SERIES

Exponential Theorem: For all value of x, positive or negative, integral or fractional, real or complex numbers.

$$e^x = 1 + x + \frac{x^2}{2!} + \frac{x^3}{3!} + \frac{x^4}{4!} + ... \qquad ...(i)$$

The series on the right hand side of the above relation is called the exponential series.

Some Important Results of Exponential Series e^x:

(i) When $x = 0$, the series of eqn. *(i)* becomes

$$e^0 = 1 + 0 + 0 + ... = 1$$

(ii) When $x = 1$, then series of eqn. *(i)* becomes

$$e^1 = e = 1 + \frac{1}{1!} + \frac{1}{2!} + \frac{1}{3!} + ...$$

and value of e lies between 2 and 3.

i.e., $2 < e < 3$.

(iii) When $x = 2$, the series of eqn. (*i*) becomes

$$e^2 = 1 + 2 + \frac{2^2}{2!} + \frac{2^3}{3!} + \ldots$$

This means that sum of the series on the right hand side is same as the square of the number e.

(iv) When $x = -1$, then

$$e^{-1} = \frac{1}{e} = 1 - \frac{1}{1!} + \frac{1}{2!} - \frac{1}{3!} + \ldots.$$

$$= \frac{1}{2!} - \frac{1}{3!} + \frac{1}{4!} - \frac{1}{5!} + \ldots.$$

The sum of this series is the reciprocal of e.

(v) When $x = -y$, then

$$e^{-y} = 1 - \frac{y}{1!} + \frac{y^2}{2!} - \frac{y^3}{3!} + \ldots$$

or $$e^{-x} = 1 - \frac{x}{1!} + \frac{x^2}{2!} - \frac{x^3}{3!} + \ldots \quad (\because y = x)$$

(vi) $e^x + e^{-x} = 2\left[1+\frac{x^2}{2!}+\frac{x^4}{4!}+....\right]$

(vii) $e^x - e^{-x} = 2\left[\frac{x}{1!}-\frac{x^3}{3!}+\frac{x^5}{5!}+...\right]$

(viii) $\cos\theta = 1-\frac{\theta^2}{2!}+\frac{\theta^4}{4!}+....$

$\sin\theta = \theta-\frac{\theta^3}{3!}+\frac{\theta^5}{5!}-....$

Exponential Theorem:

$a^x = e^{x\log_e a}$

$$= 1+\frac{x\log_e a}{1!}+\frac{x^2(\log_e a)^2}{2!}+\frac{x^3(\log_e a)^3}{3!}+...$$

(for $a > 0$)

The Valve of e: The value of e lies between 2 and 3 *i.e.*, $2 < e < 3$, and hence $e = \left(1+\frac{1}{n}\right)^n$ as $n \to \infty$

The value of e upto ten places of decimals is found as

$$e = 2.7182818284.$$

$$\lim_{x\to\infty}(1+1/x)^x = \lim_{x\to\infty}(1+x)^{1/x} = e$$

Logarithmic Series: If $-1 < x < 1$, we have

$$\log(1+x) = x - \frac{x^2}{2} + \frac{x^3}{3} - \frac{x^5}{5} + \ldots$$

This is known as Logarithmic series.

Some Important Deductions:

(*i*) $\log(1+x) = x - \frac{x^2}{2} + \frac{x^3}{3} - \frac{x^4}{4} + \ldots$

where $-1 < x \leq 1$

(*ii*) $\log(1-x) = -\left(x + \frac{x^2}{2} + \frac{x^3}{3} + \ldots\right)$

where $-1 \leq x < 1$

(*iii*) $\log(1+x) - \log(1-x) = 2\left(x + \frac{x^3}{3} + \frac{x^5}{5} + \ldots\right)$

(*iv*) $\log\left(\frac{1+x}{1-x}\right) = 2\left(x + \frac{x^3}{3} + \frac{x^5}{5} + \ldots\right);$

where $-1 < x < 1$

(*v*) $\log 2 = 1 - \frac{1}{2} + \frac{1}{3} - \frac{1}{4} + \ldots$ up to ∞

Properties of Logarithms: In performing, expansions are make use of the following properties of logarithms.

(i) $\log xy = \log x + \log y$

(ii) $\log \frac{x}{y} = \log x - \log y$

(iii) $\log x^y = y \log x$

(iv) $a^x = e^{x \log a}$

(v) $\log 1 = 0.$

The Difference between the exponential and logarithmic series:

(i) The exponential series is valid for all values of x. The logarithmic series is valid when $|x| < 1$, *i.e.*, $-1 < x < 1$.

(ii) In exponential series, the denominator of the terms involve the factorial, whereas in logarithmic series the factorial does not occur.

(iii) In the exponential series e^x all the terms are positive whereas in the series

$\log (1+x) = x - \frac{x^2}{2} + \frac{x^3}{3}$ ... the terms carry

alternatively +ve and -ve sign.

8

TRIGONOMETRIC FUNCTIONS

Trigonometry: The word trigonometry is derived from two Greek words, trigonon (triangle) and metron (to measure). Therefore, literal meaning of Trigonometry is "to measure a triangle". But now a days it is defined as that branch of mathematics which deals with angles, whether of a triangle or any other figure.

An Angle: In trigonometry an angle is defined as the amount of rotation made by a straight line from one position to another position about a point. If the initial side OX moves in anticlockwise direction to the terminal side OP from the vertex O, then

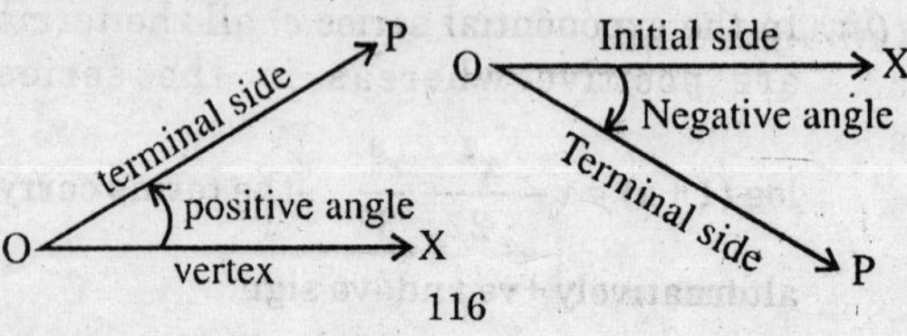

the angle XOP as shown is called a positive angle. But if, on the other hand, the initial side OX moves in the clockwise direction as shown in figure, then XOP, traced out in this manner, is called a negative angle.

Measurement of Angles: In general, the angles are measured in degrees or radians which are defined as follows:

Degrees: A right angle is divided into 90 equal parts and each part is called a degree. Thus a right angle is equal to 90 degrees. One degree is denoted by 1°. A degree is divided into sixty equal parts and each part is called a minute and is denoted by 1′ : A minute is divided into sixty equal parts and each part is called a second and is denoted by 1″.

Thus we have,

1 right angle = 90° (read as 90 degrees)

1° = 60′ (read as 60 minutes)

1′ = 60″ (read as 60 seconds)

Radians: A radian is the angle subtended at the centre of a circle by an arc equal in length to the radius of the circle.

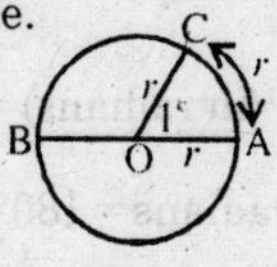

C r r 1ᶜ B O r A

In this figure OA = OC = arc AC = r = radius of the circle, then measurement of $\angle$ AOC is one radian and is denoted by 1^c. Thus $\angle AOC = 1^c$

A Constant Number π: The ratio of the circumference to the diameter of a circle is always equal to a constant and this constant is denoted by the Greek letter π.

Thus π = Circumference/diameter.

∴ If r is the radius of a circle, then its circumference = $2\pi r$.

The constant π is an irrational number and its approximate value is taken as $\frac{22}{7}$.

Relation between an Arc and an Angle: If s is the length of an arc of a circle of radius r, then the angle θ (in radians) subtended by this arc at the centre of the circle is given by

$\theta = s/r$ or $s = r\theta$

i.e., arc = radius × angle in radians.

Thus, from the above figure.

$$\angle AOB = \frac{\text{arc ACB}}{r} \text{ (in radians)} = \frac{\pi r}{r} = \pi \text{ radians}$$

Hence, we have, π radians = 180° = 2 right angles.

or 1 radian = $\frac{180}{\pi}$ degrees = $\frac{180}{22} \times 7$ degrees

= 57°17′44.8″ (Appr.)

Sectorial Area: Let OAB be a sector having central angle θ^c and radius r. Then area of the sector OAB is given by $\frac{1}{2}r^2\theta$.

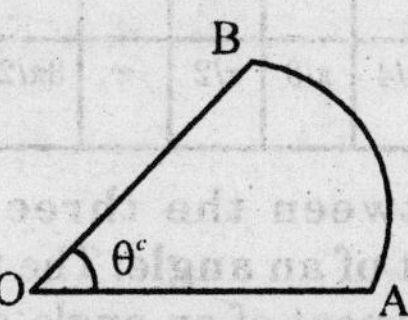

Quadrants: Let XOX′ and YOY′ be two mutually perpendicular lines in any plane. These lines divide the plane into four parts and each one of them is called quadrant.

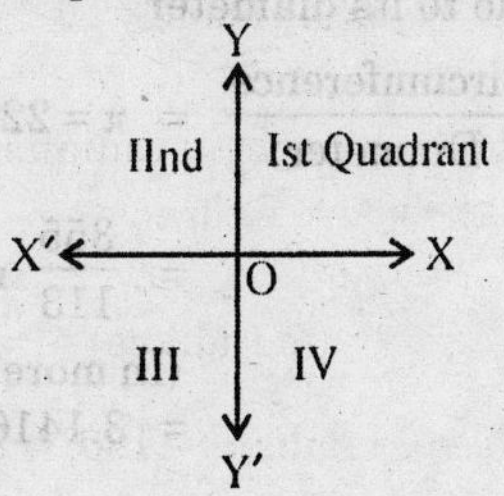

(i) The region XOY is called First Quadrant.

(ii) The region YOX′ is called the Second Quadrant.

(iii) The region X′OY′ is called Third Quadrant.

(iv) The region Y′OX is called the Fourth Quadrant.

Radians measure of Some Common angles:

Angle in Degrees	30°	45°	60°	90°	180°	270°	360°	540°	720°
Angle in Radian	$\pi/6$	$\pi/4$	$\pi/3$	$\pi/2$	π	$3\pi/2$	2π	3π	4π

Relation between the three systems of measurement of an angle: The three systems of the measurement of an angle are related by formula

$$\pi \text{ radian} = 180° = 200^g.$$

Theorem: Circumference of a cirlce bears a constant ratio to its diameter.

$$i.e., \quad \frac{\text{Circumference}}{\text{Diameter}} = \pi = 22/7$$

$$= \frac{355}{113} \text{ nearly}$$

(In more accurately)

$= 3.1416$ nearly.

Theorem: The angle, in radians, subtended by an arc of the circle at the centre $= \dfrac{\text{arc}}{\text{radius}}$

i.e., θ = angle subtended at the centre.

$$= \frac{\text{Length of arc}}{\text{Radius of the circle}}$$

$$= \frac{AC}{AB}$$

Trigonometric Ratios or functions: Let the revolving line OP start from its initial position OX and trace out an angle XOP = θ in any one of the four quadrants. From P draw PM perpendicular to X'OX.

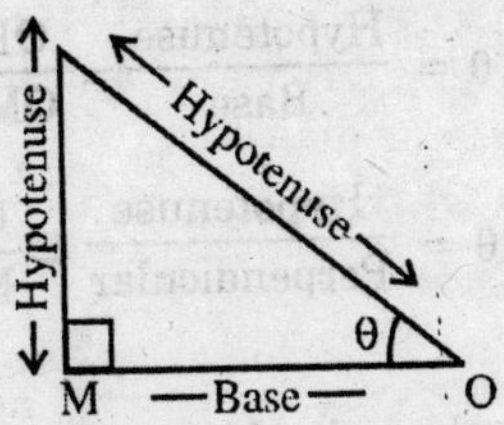

Now in the right angled triangle POM, if θ is the angle of reference, then MP, the side opposite to θ

is called perpendicular, OP the side opposite to right angle is called the hypotenuse and OM, the third side is called base.

(i) $\sin\theta = \dfrac{\text{Perpendicular}}{\text{Hypotenuse}} = \dfrac{MP}{OP}$

(ii) $\cos\theta = \dfrac{\text{Base}}{\text{Hypotenuse}} = \dfrac{OM}{OP}$

(iii) $\tan\theta = \dfrac{\text{Perpendicular}}{\text{Base}} = \dfrac{MP}{OM}$

(iv) $\cot\theta = \dfrac{\text{Base}}{\text{Perpendicular}} = \dfrac{OM}{MP}$

(v) $\sec\theta = \dfrac{\text{Hypotenuse}}{\text{Base}} = \dfrac{OP}{OM}$

(vi) $\text{cosec}\,\theta = \dfrac{\text{Hypotenuse}}{\text{Perpendicular}} = \dfrac{OP}{MP}$

Remember

(a) The student should not commit mistake by regarding $\sin\theta$ as $\sin x\,\theta$. $\sin\theta$ is correctly read as the sine of the angle θ.

(b) The student should note that $(\sin\theta)^n$ is written as $\sin^n\theta$ if $n \neq -1$.

For Example,

$$(\sin\theta)^2 = \sin^2\theta$$
$$(\sin\theta)^3 = \sin^3\theta$$

but $(\sin\theta)^{-1} \neq \sin^{-1}\theta$; Also, $\sin^{-1}\theta \neq \dfrac{1}{\sin\theta}$

Relationship between trigonometric functions:

(i) Reciprocal Relation: The following relations are obvious from the definition of t-ratios.

(a) $\sin\theta = \dfrac{1}{\text{cosec}\,\theta}$ and $\text{cosec}\,\theta = \dfrac{1}{\sin\theta}$

(b) $\cos\theta = \dfrac{1}{\sec\theta}$ and $\sec\theta = \dfrac{1}{\cos\theta}$

(c) $\tan\theta = \dfrac{1}{\cot\theta}$ and $\cot\theta = \dfrac{1}{\tan\theta}$

(d) $\sin\theta \times \text{cosec}\,\theta = 1$

(e) $\cos\theta \times \sec\theta = 1$

(f) $\tan\theta \times \cot\theta = 1$

(ii) Quotient Relation

(a) $\tan\theta = \dfrac{\sin\theta}{\cos\theta}$

(b) $\cot\theta = \dfrac{\cos\theta}{\sin\theta}$

(iii) Square Relations

(a) $\sin^2\theta + \cos^2\theta = 1$

(b) $\sec^2\theta - \tan^2\theta = 1$

(c) $\text{cosec}^2\theta - \cot^2\theta = 1$

Trigonometric Identities: Trigonometric Identities is a statement of equality between the expressions and is true for all values of the variable involved. In order to prove them, we give below some of the methods to be used:

Method I: Simplify L.H.S. or R.H.S. which ever is complicated and prove it to be equal to the other side.

Illustrations: Prove that

$$\sin^4\theta + \cos^4\theta = 1 - 2\sin^2\theta\cos^2\theta$$

Soln:

$$\begin{aligned} \text{L.H.S.} &= \sin^4\theta + \cos^4\theta \\ &= (\sin^2\theta)^2 + (\cos^2\theta)^2 \\ &= (\sin^2\theta + \cos^2\theta)^2 - 2\sin^2\theta\cos^2\theta \\ &\quad [\because a^2 + b^2 = (a+b)^2 - 2ab] \\ &= (1)^2 - 2\sin^2\theta\cos^2\theta \\ &= 1 - 2\sin^2\theta\cos^2\theta \\ &= \text{R.H.S.} \end{aligned}$$

Method II: Change all the trigonometrical ratios in terms of the sines and cosines of the angles.

Illustration: Prove that

$$(\tan\alpha + \cot\alpha)^2 = \sec^2\alpha \operatorname{cosec}^2\alpha$$

Soln:

$$\text{L.H.S.} = (\tan\alpha + \cot\alpha)^2$$

$$= \left(\frac{\sin\alpha}{\cos\alpha} + \frac{\cos\alpha}{\sin\alpha}\right)^2$$

$$= \left(\frac{\sin^2\alpha + \cos^2\alpha}{\cos\alpha.\sin\alpha}\right)^2$$

$$= \left(\frac{1}{\cos\alpha\sin\alpha}\right)^2$$

$$= \frac{1}{\cos^2\alpha}\cdot\frac{1}{\sin^2\alpha}$$

$$= \sec^2\alpha \,.\, \operatorname{cosec}^2\alpha$$

$$= \text{R.H.S.}$$

Method III: Simplify L.H.S. and R.H.S. both, if both are complicated and then prove L.H.S. = R.H.S.

Illustration: Prove that

$\sin^2 A \tan A + \cos^2 A \cot A + 2 \sin A \cos A$

$= \tan A + \cot A$

Soln: L.H.S. $= \sin^2 A \tan A + \cos^2 A \cot A + 2 \sin A \cos A$

$$= \sin^2 A \frac{\sin A}{\cos A} + \cos^2 A \frac{\cos A}{\sin A} + 2 \sin A \cos A$$

$$= \frac{\sin^3 A}{\cos A} + \frac{\cos^3 A}{\sin A} + 2 \sin A \cos A$$

$$= \frac{\sin^4 A + \cos^4 A + 2\sin^2 A \cos^2 A}{\cos A \sin A}$$

$$= \frac{(\sin^2 A + \cos^2 A)^2}{\cos A \sin A}$$

$[\because (a+b)^2 = a^2 + b^2 + 2ab]$

$$= \frac{1}{\cos A \sin A}$$

Now R.H.S. $= \tan A + \cot A = \frac{\sin A}{\cos A} + \frac{\cos A}{\sin A}$

$$= \frac{\sin^2 A + \cos^2 A}{\cos A \sin A}$$

$$= \frac{1}{\cos A \sin A}$$

Hence, L.H.S. = R.H.S. proved.

Method IV: From the identity to be proved, obtain another identity which is obviously true by cross-multiplication or by transpositioning.

Illustration: Prove that

$$\frac{1}{\sec A + \tan A} - \frac{1}{\cos A} = \frac{1}{\cos A} - \frac{1}{\sec A - \tan A}$$

Soln.: $\frac{1}{\sec A + \tan A} - \frac{1}{\cos A}$

$$= \frac{1}{\cos A} - \frac{1}{\sec A - \tan A}$$

If transposing

$$\frac{1}{\sec A + \tan A} + \frac{1}{\sec A - \tan A} = \frac{1}{\cos A} + \frac{1}{\cos A}$$

$$\Rightarrow \frac{\sec A - \tan A + \sec A + \tan A}{(\sec A + \tan A)(\sec A - \tan A)} = \frac{2}{\cos A}$$

$$\Rightarrow \frac{2\sec A}{\sec^2 A - \tan^2 A} = \frac{2}{\cos A}$$

$$\Rightarrow \frac{2\sec A}{1} = \frac{2}{\cos A}$$

$$\Rightarrow \frac{2}{\cos A} = \frac{2}{\cos A}$$ which is true.

Sign of trigonometric ratios in different quadrants:

(i) In first quadrant all the t-ratios are positive

(ii) In second quadrant sin θ and cosec θ are +ve and other t-ratios are –ve.

(iii) In third quadrant, tan θ and cot θ are +ve and other t-ratios are –ve.

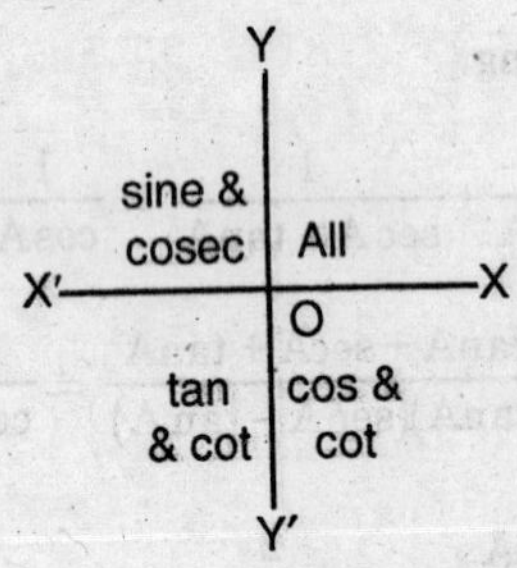

(iv) In fourth quadrant, cos θ and sec θ are +ve and other t-ratios are –ve.

Limits to the values of t-ratios:

(*i*) sin θ and cos θ can not be numerically greater than 1.

(*ii*) sec θ and cosec θ can not be numerically less than 1.

(*iii*) tan θ and cot θ can have any numerical values.

Trigonometric Ratios of Standard Angles

	0°	15°	18°	22.5°	30°	36°	45°	60°	67.5°	90°
sin	0	$\frac{\sqrt{6}-\sqrt{2}}{4}$	$\frac{\sqrt{5}-1}{4}$	$\frac{\sqrt{2-\sqrt{2}}}{2}$	$\frac{1}{2}$	$\frac{\sqrt{10-2\sqrt{5}}}{4}$	$\frac{1}{\sqrt{2}}$	$\frac{\sqrt{3}}{2}$	$\frac{\sqrt{\sqrt{2}+1}}{\sqrt{(2\sqrt{2})}}$	1
cos	1	$\frac{\sqrt{6}+\sqrt{2}}{4}$	$\frac{\sqrt{10+2\sqrt{5}}}{4}$	$\frac{\sqrt{\sqrt{2}+1}}{(2\sqrt{2})}$	$\frac{\sqrt{3}}{2}$	$\frac{\sqrt{5}+1}{4}$	$\frac{1}{\sqrt{2}}$	$\frac{1}{2}$	$\frac{\sqrt{2-\sqrt{2}}}{2}$	0
tan	0	$2-\sqrt{3}$	$\frac{\sqrt{25-10\sqrt{5}}}{5}$	$\sqrt{2}-1$	$\frac{1}{\sqrt{3}}$	$\sqrt{5-2\sqrt{5}}$	1	$\sqrt{3}$	$\sqrt{2}+1$	not defined

Trigonometric Ratios of Allied Angles

Allied angle: Two angles are said to be allied when their sum or difference is a multiple of 90°. The angle −θ, 90° ± θ, 180° ± θ, etc. are angles allied to the angle θ.

t-ratio	$-\theta$	$90°-\theta$	$90°+\theta$	$180°-\theta$	$180°+\theta$	$270°-\theta$	$270°+\theta$	$360°-\theta$	$360°+\theta$
$\sin\theta$	$-\sin\theta$	$\cos\theta$	$\cos\theta$	$\sin\theta$	$-\sin\theta$	$-\cos\theta$	$-\cos\theta$	$-\sin\theta$	$\sin\theta$
$\cos\theta$	$\cos\theta$	$\sin\theta$	$-\sin\theta$	$-\cos\theta$	$-\cos\theta$	$-\sin\theta$	$\sin\theta$	$\cos\theta$	$\cos\theta$
$\tan\theta$	$-\tan\theta$	$\cot\theta$	$-\cot\theta$	$-\tan\theta$	$\tan\theta$	$\cot\theta$	$-\cot\theta$	$-\tan\theta$	$\tan\theta$
$\cot\theta$	$-\cot\theta$	$\tan\theta$	$-\tan\theta$	$-\cot\theta$	$\cot\theta$	$\tan\theta$	$-\tan\theta$	$-\cot\theta$	$\cot\theta$
$\sec\theta$	$\sec\theta$	$\text{cosec}\,\theta$	$-\text{cosec}\,\theta$	$-\sec\theta$	$-\sec\theta$	$-\text{cosec}\,\theta$	$\text{cosec}\,\theta$	$\sec\theta$	$\sec\theta$
$\text{cosec}\,\theta$	$-\text{cosec}\,\theta$	$\sec\theta$	$\sec\theta$	$\text{cosec}\,\theta$	$-\text{cosec}\,\theta$	$-\sec\theta$	$-\sec\theta$	$-\text{cosec}\,\theta$	$\text{cosec}\,\theta$

The above results may be obtained by the following rules:

Rule 1: The trigonometrical ratio of $90° \pm \theta$, $270° \pm \theta$ is changed *i.e.,* sin → cos, cos → sin, tan → cot. The positive or negative sign depends on quadrant.

For example, to get a value of $\sin(270° + \theta)$, sin is changed to cos, and since angle $270° + \theta$ is in the 4th quadrant in which sign of sin is –ve.

$\therefore \sin(270° + \theta) = -\cos\theta$.

Rule 2: The trigonometrical ratio of $180° \pm \theta$, $360° \pm \theta$ is not changed and the +ve or –ve sign depend on the quadrant rule.

For example, to get the value of $\cos(180° - \theta)$, cos is not changed, and since angle $180° - \theta$ is in the second quadrant in which the sign of cos is –ve.

$\therefore \cos(180° - \theta) = -\cos\theta$.

Hence, in general consider the angles $\frac{1}{2} n\pi + \theta$ and $\frac{1}{2} n\pi - \theta$, $n \in I$, then

(*i*) assuming that $0 < \theta < 90°$, the result has the plus or the minus sign according as the given function is positive or negative in that quadrant.

(ii) If n is even, the result contains the same trigonometric function as the given expression, but if n is odd, the result contains the corresponding co-function, *i.e.*, sine becomes cosine, tangent becomes cotangent, secant becomes cosecant and vice-versa.

For example, consider $\cos(450° - \theta)$. We have $450° = 5 \times 90°$, so $450° - \theta$ is a first quadrant angle, and n is odd, so $\cos(450° - \theta) = \sin\theta$

Also, this can be found as

$\cos(450° - \theta) = \cos(360° + 90° - \theta) = \cos(90° - \theta)$
$= \sin\theta$

Some interesting results about allied angles:

(1) $\cos n\pi = (-1)^n$, $\sin n\pi = 0$

(2) $\cos(n\pi + \theta) = (-1)^n \cos\theta$

$\sin(n\pi + \theta) = (-1)^n \sin\theta$

(3) $\cos\left(\frac{n\pi}{2} + \theta\right) = (-1)^{\frac{n+1}{2}} \sin\theta$ if n is odd

$= (-1)^{n/2} \cos\theta$ if n is even.

(4) $\sin\left(\frac{n\pi}{2} + \theta\right) = (-1)^{\frac{n-1}{2}} \cos\theta$ if n is odd

$= (-1)^{n/2} \sin\theta$ if n is even.

n	$\frac{n\pi}{2}+\theta$	$\cos\left(\frac{n\pi}{2}+\theta\right)$
1	$\frac{\pi}{2}+\theta$	$-\sin\theta$
2	$\pi+\theta$	$-\cos\theta$
3	$\frac{3\pi}{2}+\theta$	$\sin\theta$
4	$2\pi+\theta$	$\cos\theta$
–	–	–

Periodicity and Graphical representation of trigonometric functions: If a function $f(x) = f(x + \alpha)$, where α is the least positive constant, then $f(x)$ is called the periodic function of x and α is called its period.

Theorem: $\sin\theta$, $\cos\theta$, $\sec\theta$, $\operatorname{cosec}\theta$ are periodic functions with period 2π whereas $\tan\theta$ and $\cot\theta$ are periodic functions with period π.

i.e., $\sin(\theta + 2\pi) = \sin\theta$

$\cos(\theta + 2\pi) = \cos\theta$

$\sec(\theta + 2\pi) = \sec\theta$

$\operatorname{cosec}(\theta + 2\pi) = \operatorname{cosec}\theta$

i.e., sinθ, cosθ, secθ, cosecθ remains unchanged, when θ is increased by the least positive constant 2π.

Hence, sinθ, cosθ, secθ, cosecθ are periodic function with period 2π...

Again $\tan(\theta + \pi) = \tan\theta$; $\cot(\theta + \pi) = \cot\theta$

i.e., tanθ, cotθ remain unchanged when θ is changed to θ + π, where π is the least positive constant and is called period.

Graph of sin *x*. Table of Values

x	0°	30°	60°	90°	120°	150°	180°	210°	240°	270°	300°	330°	360°
$y = \sin x$	0	.5	.87	1	.87	.5	0	–.5	–.87	–1	–.87	–.5	0

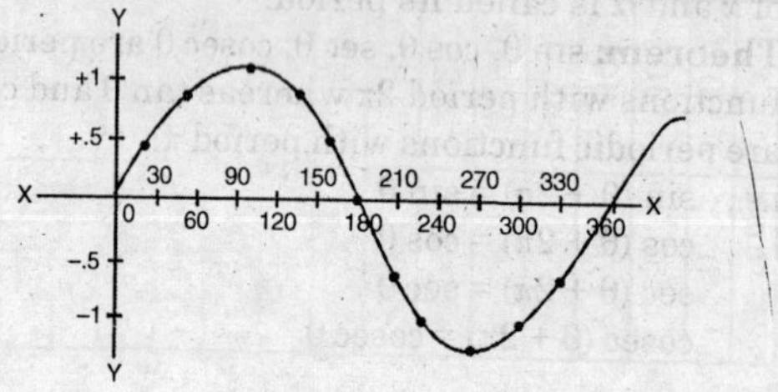

Graph of cos x : Table of Values

x	0°	30°	60°	90°	120°	150°	180°	210°	240°	270°	300°	330°	360°
$y = \cos x$	1	.87	.5	0	–.5	–.87	–1	–.87	–.5	0	.5	.87	1

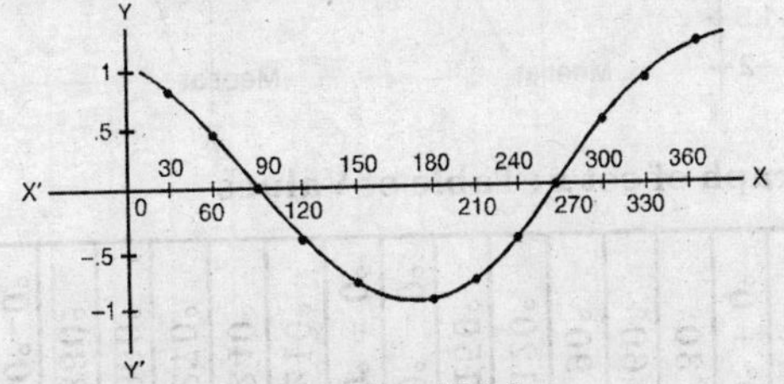

Graph of tan x : Table of Values

x	0°	30°	60°	90–0°	90+0°	120°	150°	180°	210°	240°	270–0°	270+0°	300°	330°	360°
$y = \tan x$	0	.58	1.73	$+\infty$	$-\infty$	–1.73	–.58	0	.58	1.73	$+\infty$	$-\infty$	–1.73	–.58	0

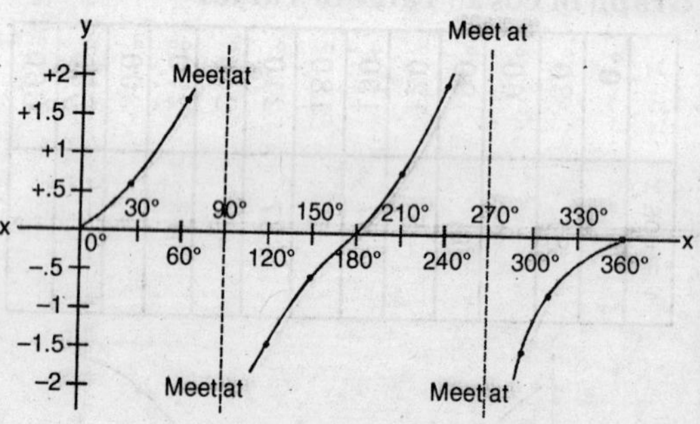

Graph of cot x : Table of Values

x	0° + 0°	30°	60°	90°	120°	150°	180° − 0°	180° + 0°	210°	240°	270°	300°	330°	360°−0°
$y = \cot x$	∞	1.73	.58	0	−.58	−1.73	− ∞	+ ∞	1.73	.58	0	−.58	−1.73	−∞

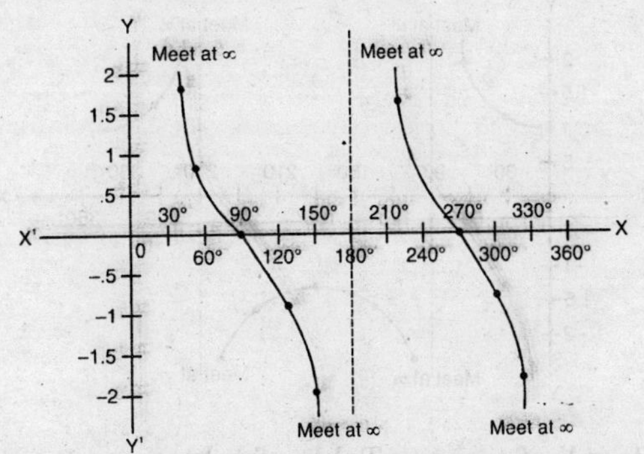

Graph of sec x : Table of Values

x	0°	30°	60°	90° − 0°	90° + 0°	120°	150°	180°	210°	240°	270° − 0°	270° + 0°	300°	330°	360°
$y = \sec x$	1	1.15	2	+∞	− ∞	− 2	− 1.15	−1	−1.15	−2	−∞	+ ∞	2	1.15	1

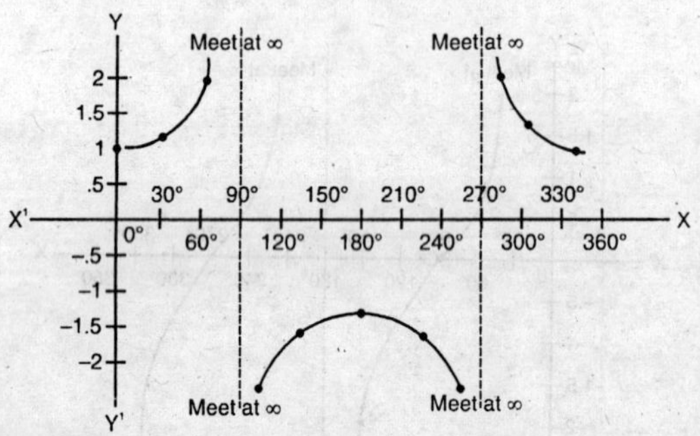

Graph of cosec x : Table of values

x	$y = \text{cosec } x$
0° + 0°	+ ∞
30°	2
60°	1.15
90°	1
120°	1.15
150°	2
180° − 0°	+ ∞
180° + 0°	− ∞
210°	−2
240°	−1.15
270°	−1
300°	−1.15
330°	−2
360° − 0°	−∞

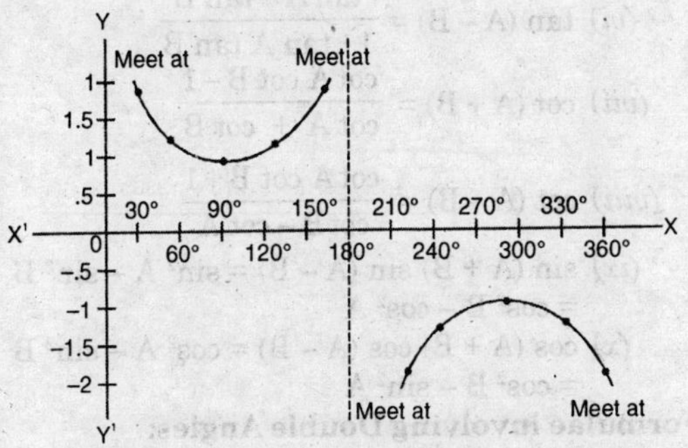

Trigonometrical Ratio of Compound Angle:

Compound Angle: An angle made up by the sum or difference of two or more angles is called compound angle.

I. Addition and Subtraction Formulae

(i) $\sin(A + B) = \sin A \cos B + \cos A \sin B$

(ii) $\sin(A - B) = \sin A \cos B - \cos A \sin B$

(iii) $\cos(A + B) = \cos A \cos B - \sin A \sin B$

(iv) $\cos(A - B) = \cos A \cos B + \sin A \sin B$

(v) $\tan(A + B) = \dfrac{\tan A + \tan B}{1 - \tan A \tan B}$

(*vi*) $\tan(A-B) = \dfrac{\tan A - \tan B}{1+\tan A \tan B}$

(*vii*) $\cot(A+B) = \dfrac{\cot A \cot B - 1}{\cot A + \cot B}$

(*viii*) $\cot(A-B) = \dfrac{\cot A \cot B + 1}{\cot B - \cot A}$

(*ix*) $\sin(A+B)\sin(A-B) = \sin^2 A - \sin^2 B$
$= \cos^2 B - \cos^2 A$

(*x*) $\cos(A+B)\cos(A-B) = \cos^2 A - \sin^2 B$
$= \cos^2 B - \sin^2 A$

Formulae involving Double Angles:

Replacing $A = B = \theta$ in above formulae, we get

(*i*) $\sin 2\theta = 2\sin\theta\cos\theta = \dfrac{2\tan\theta}{1+\tan^2\theta}$

(*ii*) $\cos 2\theta = \cos^2\theta - \sin^2\theta = 1 - 2\sin^2\theta$

$= 2\cos^2\theta - 1 = \dfrac{1-\tan^2\theta}{1+\tan^2\theta}$

(*iii*) $1 + \cos 2\theta = 2\cos^2\theta$; $1 - \cos 2\theta = 2\sin^2\theta$;

$\cos^2\theta = \dfrac{1}{2}(1+\cos 2\theta)$;

$\sin^2\theta = \dfrac{1}{2}(1-\cos 2\theta)$

(iv) $\tan 2\theta = \dfrac{2\tan\theta}{1-\tan^2\theta}$

Formulae Involving Half Angles:

Replacing 2θ by θ in above formulae, we get

(i) $\sin\theta = 2\sin\dfrac{\theta}{2}\cos\dfrac{\theta}{2} = \dfrac{2\tan\dfrac{\theta}{2}}{1+\tan^2\dfrac{\theta}{2}}$

(ii) $\cos\theta = \cos^2\dfrac{\theta}{2} - \sin^2\dfrac{\theta}{2}$

$= 1 - 2\sin^2\dfrac{\theta}{2} = 2\cos^2\dfrac{\theta}{2} - 1$

$= \dfrac{1-\tan^2\dfrac{\theta}{2}}{1+\tan^2\dfrac{\theta}{2}}$

(iii) $\tan\theta = \dfrac{2\tan\dfrac{\theta}{2}}{1-\tan^2\dfrac{\theta}{2}}$

Formulae involving Triple Angles

(i) $\sin 3\theta = 3\sin\theta - 4\sin^3\theta$

(ii) $\cos 3\theta = 4\cos^3\theta - 3\cos\theta$

(iii) $\tan 3\theta = \dfrac{3\tan\theta - \tan^3\theta}{1 - 3\tan^2\theta}$

(iv) $\cot 3\theta = \dfrac{\cot^3\theta - 3\cot\theta}{3\cot^2\theta - 1}$

Formulae for changing the Product into Sum or Difference.

$$\left.\begin{aligned} 2\sin A\cos B &= \sin(A+B) + \sin(A-B) \\ 2\cos A\sin B &= \sin(A+B) - \sin(A-B) \\ 2\cos A\cos B &= \cos(A+B) + \cos(A-B) \end{aligned}\right\}, A > B$$

$$2\sin A\sin B = \cos(A-B) - \cos(A+B)$$

Formulae for changing the Sum or Difference into Product

Substituting $A + B = C$, $A - B = D$ and hence $A = \dfrac{(C+D)}{2}$, $B = \dfrac{(C-D)}{2}$ in the above formulae, then

(i) $\sin C + \sin D = 2\sin\dfrac{C+D}{2}\cos\dfrac{C-D}{2}$

(ii) $\sin C - \sin D = 2\cos\dfrac{C+D}{2}\sin\dfrac{C-D}{2}$

(iii) $\cos C + \cos D = 2\cos\frac{C+D}{2}\cos\frac{C-D}{2}$

(iv) $\cos C - \cos D = 2\sin\frac{C+D}{2}\sin\frac{D-C}{2}$

Formulae involving Three or More Angles

(i) $\sin(\alpha+\beta+\gamma) = \sin\alpha\cos\beta\cos\gamma + \cos\alpha\sin\beta\cos\gamma + \cos\alpha\cos\beta\sin\gamma - \sin\alpha\sin\beta\sin\gamma$

(ii) $\cos(\alpha+\beta+\gamma) = \cos\alpha\cos\beta\cos\gamma - \cos\alpha\sin\beta\sin\gamma - \sin\alpha\cos\beta\sin\gamma - \sin\alpha\sin\beta\cos\gamma$

(iii) $\tan(\alpha+\beta+\gamma)$

$$= \frac{\tan\alpha+\tan\beta+\tan\gamma-\tan\alpha\tan\beta\tan\gamma}{1-(\tan\alpha\tan\beta+\tan\beta\tan\gamma+\tan\gamma\tan\alpha)}$$

$$= \frac{\sigma_1-\sigma_3}{1-\sigma_2}$$

(iv) $\sin(A_1+A_2+....+A_n) = \cos A_1\cos A_2 ... \cos A_n(\sigma_1-\sigma_3+\sigma_5-....)$

(v) $\cos(A_1+A_2+...+A_n) = \cos A_1\cos A_2 ... \cos A_n(1-\sigma_2+\sigma_4-\sigma_6+...)$

(vi) $\tan(A_1 + A_2 + \ldots + A_n)$

$$= \frac{\sigma_1 - \sigma_3 + \sigma_5 - \ldots}{1 - \sigma_2 + \sigma_4 - \sigma_6 + \ldots}, \text{ where}$$

$\sigma_1 = \Sigma \tan A_1$, $\sigma_2 = \Sigma \tan A_1 \tan A_2$,
$\sigma_3 = \Sigma \tan A_1 \tan A_2 \tan A_3$ etc.

Polar form and Extremas:

Polar form for $a \cos\theta + b \sin\theta$: Let $x = a\cos\theta + b\sin\theta$. Then x can be converted into cos or into sin as follows:

(i) Put $a = r\cos\alpha$, $b = r\sin\alpha$. Then

$r = \sqrt{a^2 + b^2}$ and $\tan\alpha = b/a$.

Note that α should be evaluated by $\cos\alpha = a/r$, $\sin\alpha = b/r$. Then we get $x = r\cos(\theta - \alpha)$.

(ii) By putting $a = r\sin\alpha$, $b = r\cos\alpha$,

We get, $x = r\sin(\theta + \alpha)$,

$$r = \sqrt{(a^2 + b^2)} \text{ and } \cos\alpha = b/r$$

$\sin\alpha = a/r$

For example, Let $x = -\sqrt{3}\cos\theta + \sin\theta$

putting, $-\sqrt{3} = r\cos\alpha$, $1 = r\sin\alpha$, we get

$r = 2$, $\cos\alpha = -\sqrt{3}/2$, $\sin\alpha = 1/2$, so $\alpha = 2\pi/3$.

$\therefore x = 2\cos(\theta - 2\pi/3)$.

dentity: A trigonometric equation is an identity f it is true for all values of the angles involved.

Conditional Identity: A conditional identity is n identity which holds if the variables satisfy a given condition. When three angles A, B, C are such that $A + B + C = 180°$ (or that A, B, C are the angles of a triangle), several identities hold between the trigonometrical function A, B, C or their multiples and sub-multiples.

Some Important Identities:

If $A + B + C = \pi$, then

(i) $\tan A + \tan B + \tan C = \tan A \tan B \tan C$

(ii) $\cot A \cot B + \cot B \cot C + \cot C \cot A = 1$

(iii) $\sin 2A + \sin 2B + \sin 2C = 4 \sin A \sin B \sin C$

(iv) $\cos 2A + \cos 2B + \cos 2C = -1 - 4 \cos A \cos B \cos C$

(v) $\cos^2 A + \cos^2 B + \cos^2 C = 1 - 2 \cos A \cos B \cos C$

(vi) $\cos A + \cos B + \cos C = 1 + 4 \sin \frac{A}{2} \sin \frac{B}{2} \sin \frac{C}{2}$

(vii) $\tan \frac{A}{2} \tan \frac{B}{2} + \tan \frac{B}{2} \tan \frac{C}{2} + \tan \frac{C}{2} \tan \frac{A}{2} = 1$

(*viii*) $\cot \frac{A}{2} + \cot \frac{B}{2} + \cot \frac{C}{2} = \cot \frac{A}{2} \cot \frac{B}{2} \cot \frac{C}{2}$

(*ix*) $\cos \alpha + \cos \beta + \cos \gamma + \cos (\alpha + \beta + \gamma)$

$$= 4 \cos \left(\frac{\alpha+\beta}{2}\right) \cos \left(\frac{\beta+\gamma}{2}\right) \cos \left(\frac{\gamma+\alpha}{2}\right)$$

(*x*) $\sin \alpha + \sin \beta + \sin \gamma - \sin (\alpha + \beta + \gamma)$

$$= 4 \sin \left(\frac{\alpha+\beta}{2}\right) \sin \left(\frac{\beta+\gamma}{2}\right) \sin \left(\frac{\gamma+\alpha}{2}\right)$$

Trigonometric Equations: Equation involving one or more than one trigonometric ratios of unknown angles are called trigonometric equations.

For example, (*i*) $2 \sin \theta + \cos 2\theta = 0$

(*ii*) $3 \sin^2 \theta - 4 \cos^2 \theta = \sin \theta$ are the trigonometric equations in unknown angle θ. A value of unknown angle satisfying the given equation is called the solution of the equation. We mainly consider the three types of equations:

(*i*) One equation in one variable

(*ii*) Two equations in one variable

(*iii*) Two equations in two variables.

General Solution of one trigonometric equation in one variable.

1. If $\sin\theta = \sin\alpha$ or $\operatorname{cosec}\theta = \operatorname{cosec}\alpha$, then
$$\theta = n\pi + (-1)^n \alpha,\ n \in I$$
2. If $\cos\theta = \cos\alpha$ or $\sec\theta = \sec\alpha$, then
$$\theta = 2n\pi \pm \alpha,\ n \in I.$$
3. If $\tan\theta = \tan\alpha$ or $\cot\theta = \cot\alpha$, then
$$\theta = n\pi + \alpha,\ n \in I.$$
4. If $\sin^2\theta = \sin^2\alpha$ or $\cos^2\theta = \cos^2\alpha$ or $\tan^2\theta = \tan^2\alpha$ etc. then $\theta = n\pi \pm \alpha,\ n \in I$

Important Deductions:

(i) $\cos\theta = 0$, then $\theta = 2n\pi \pm \pi/2$ or $n\pi \pm \pi/2$ or $n\pi + \pi/2$

(ii) $\cos\theta = 1$, then $\theta = 2n\pi$

(iii) $\cos\theta = -1$, then $\theta = (2n+1)\pi$

(iv) $\sin\theta = 0$, then $\theta = n\pi$

(v) $\sin\theta = 1$, then $\theta = 2n\pi + \pi/2$ or $n\pi + (-1)^n \pi/2$

(vi) $\sin\theta = -1$, then $\theta = 2n\pi - \pi/2$ or $n\pi - (-1)^n \pi/2$

9

INVERSE TRIGONOMETRIC FUNCTIONS

Inverse Trigonometric Functions: We know that the equation $x = \sin y$ ...(i) means that y is an angle whose sine is x or x is the sine of y.

After solving equation (i) for y, we get

$$y = \sin^{-1} x \text{ or } y = \text{arc} \sin x$$

Similarly $y = \cos^{-1} x$ if $\cos y = x$ and $y = \tan^{-1} x$ if $x = \tan y$ etc. The functions $\sin^{-1} x$, $\cos^{-1} x$, $\tan^{-1} x$, $\sec^{-1} x$, $\text{cosec}^{-1} x$ and $\cot^{-1} x$ are called inverse trigonometric functions. It is important to note that

(i) $\sin y$ is a number whereas $\sin^{-1} x$ is an angle

(ii) $\sin^{-1} \neq (\sin x)^{-1} = \dfrac{1}{\sin x}$

Properties of inverse trigonometric functions

I. Some angle can be expressed by different inverse trigonometric functions.

We know that

$$\sin 60^\circ = \frac{\sqrt{3}}{2} \Rightarrow 60^\circ = \sin^{-1}\left(\frac{\sqrt{3}}{2}\right)$$

$$\cos 60^\circ = \frac{1}{2} \Rightarrow 60^\circ = \cos^{-1}\left(\frac{1}{2}\right)$$

$$\tan 60^\circ = \sqrt{3} \Rightarrow 60^\circ = \tan^{-1}\left(\sqrt{3}\right)$$

Here, $60^\circ = \sin^{-1}\left(\frac{\sqrt{3}}{2}\right) = \cos^{-1}\left(\frac{1}{2}\right) = \tan^{-1}\sqrt{3} = \ldots$

II. Inverse property: We know that

$$x = \cos\theta, \text{ then, } \theta = \cos^{-1} x$$

$$\therefore \quad \theta = \cos^{-1}(\cos\theta) \quad (\because x = \cos\theta)$$

III. Principle of reciprocity: Following reciprocal relation exists between inverse trigonometric functions,

$$\operatorname{cosec}^{-1}\frac{1}{x} = \sin^{-1} x$$

$$\sec^{-1}\frac{1}{x} = \cos^{-1} x$$

$$\cot^{-1}\frac{1}{x} = \tan^{-1} x$$

IV. **Inverse trigonometric functions are odd functions within the principal values**

(*i*) $\sin^{-1}(-x) = -\sin^{-1} x$

(*ii*) $\text{cosec}^{-1}(-x) = -\text{cosec}^{-1} x$

(*iii*) $\tan^{-1}(-x) = -\tan^{-1} x$

V. Some fundamental formulæ

(*i*) $\sin^{-1} x + \cos^{-1} x = \dfrac{\pi}{2}$

(*ii*) $\tan^{-1}x + \cot^{-1} x = \pi/2$

(*iii*) $\text{cosec}^{-1} x + \sec^{-1} x = \pi/2$

(*iv*) $\tan^{-1} x + \tan^{-1}y = \tan^{-1}\left(\dfrac{x+y}{1-xy}\right)$

(*v*) $\tan^{-1} x - \tan^{-1} y = \tan^{-1}\left(\dfrac{x-y}{1+xy}\right)$

(*vi*) $2\tan^{-1}x = \sin^{-1}\dfrac{2x}{1+x^2}$

$$= \cos^{-1}\frac{1-x^2}{1+x^2}$$

$$= \tan^{-1}\frac{2x}{1-x^2}$$

VI. To express one inverse trigonometric function in terms of other ones:

(i) $\sin^{-1} x = \cos^{-1}\sqrt{1-x^2} = \tan^{-1}\dfrac{x}{\sqrt{1-x^2}}$

(ii) $\cos^{-1} x = \sin^{-1}\sqrt{1-x^2} = \tan^{-1}\dfrac{\sqrt{1-x^2}}{x}$

(iii) $\operatorname{cosec}^{-1}\dfrac{1}{x} = \sin^{-1}\dfrac{1}{\sqrt{1-x^2}}$

$$= \cot^{-1}\frac{\sqrt{1-x^2}}{x}$$

VII. Some Important Deductions:

(a) $\sin(\sin^{-1} x) = x\,;\ -1 \le x \le 1$

$\tan(\tan^{-1} x) = x\,;\ -\infty < x < \infty$

$\cos(\cos^{-1} x) = x\,;\ -1 \le x \le 1$

$\cot(\cot^{-1} x) = x\,;\ -\infty < x < \infty$

$\sec(\sec^{-1} x) = x\,;\ x \le -1$ or $x \ge 1$

$\operatorname{cosec}(\operatorname{cosec}^{-1} x) = x\,;\ x \le -1$ or $x \ge 1$

(b) $\sin^{-1}(\sin\theta) = \theta\,;\ -\pi/2 \le \theta \le \pi/2$

$\cos^{-1}(\cos\theta) = \theta\,;\ 0 \le \theta \le \pi$

$\tan^{-1}(\tan\theta) = \theta\ ;\ -\pi/2 < \theta < \pi/2$

$\cot^{-1}(\cot\theta) = \theta\ ;\ 0 < \theta < \pi$

$\sec^{-1}(\sec\theta) = \theta\ ;\ 0 \le \theta < \pi,\ \theta \ne \pi/2$

$\text{cosec}^{-1}(\text{cosec}\,\theta) = \theta\ ;\ -\pi/2 \le \theta \le \pi/2,\ \theta \ne$

(c) $\sin^{-1}(-x) = -\sin^{-1}x\ ;\ -1 \le x \le 1$

$\cos^{-1}(-x) = \pi - \cos^{-1}x\ ;\ -1 \le x \le 1$

$\tan^{-1}(-x) = -\tan^{-1}x\ ;\ x \in \text{R}$

$\cot^{-1}(-x) = \pi - \cot^{-1}x\ ;\ x \in \text{R}.$

$\sec^{-1}(-x) = \pi - \sec^{-1}x\ ;\ x \ge 1 \text{ or } x \le -1$

$\text{cosec}^{-1}(-x) = -\text{cosec}^{-1}x;\ x \ge 1 \text{ or } x \le -1$

10

SOLUTION OF TRIANGLES

1. **Relation between Sides and Angles of a Triangle:** In any triangle ABC, the capital letters A, B, C denote the angles and the small letters *a, b, c* denote the side opposite to the angles A, B, C respectively.

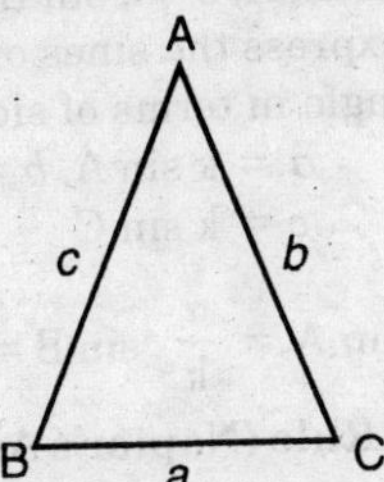

2. **Sine Rule (Law of Sines):** In any triangle

ABC, $\dfrac{a}{\sin A} = \dfrac{b}{\sin B} = \dfrac{c}{\sin C}$

[Any triangle, the sides are proportional to the sines of the opposite angles.]

The above formula can also be put as

$$\frac{\sin A}{a} = \frac{\sin B}{b} = \frac{\sin C}{c}$$

I. To express the sides of a triangle in terms of angles

From sine formula

$$\frac{a}{\sin A} = \frac{b}{\sin B} = \frac{c}{\sin C} = k \text{ (say)}$$

$\therefore \quad a = k \sin A,\ b = k \sin B,\ c = k \sin C.$

II. To express the sines of the angles of a triangle in terms of sides

$$a = k \sin A,\ b = k \sin B,$$
$$c = k \sin C$$

$$\therefore \quad \sin A = \frac{a}{k},\ \sin B = \frac{b}{k},\ \sin c = \frac{c}{k}$$

3. Tangent Rule (Napier's Analogy): In any triangle ABC, $\tan \frac{C-A}{2} = \frac{c-a}{c+a} \cot \frac{B}{2}$;

$$\tan \frac{B-C}{2} = \frac{b-c}{b+c} \cot \frac{A}{2};$$

$$\tan \frac{A-B}{2} = \frac{a-b}{a+b} \cot \frac{C}{2}$$

4. **The cosine Rule (Law of cosine):** In any triangle ABC,

$$\cos A = (b^2 + c^2 - a^2)/2bc$$
$$\cos B = (c^2 + a^2 - b^2)/2ca$$
$$\cos C = (a^2 + b^2 - c^2)/2ab$$

Here,
$$c^2 = a^2 + b^2 - 2\,ab \cos C$$
$$a^2 = b^2 + c^2 - 2bc \cos A$$
$$b^2 = c^2 + a^2 - 2ca \cos B$$

5. **Projection Rule:** In any triangle ABC

$$a = b \cos C + c \cos B$$
$$b = c \cos A + a \cos C$$
$$c = a \cos B + b \cos A$$

6. **Area of a Triangle:** The area of a ΔABC, denoted by Δ is given by the following formulae:

(i) $\Delta = \frac{1}{2}\, bc \sin A = \frac{1}{2}\, ca \sin B = \frac{1}{2}$ $ab \sin C$.

(ii) $\Delta = \sqrt{s(s-a)(s-b)(s-c)}$,

where $s = (a+b+c)/2$

(iii) $\Delta = \dfrac{b^2 \sin C \sin A}{2 \sin B} = \dfrac{c^2 \sin A \sin B}{2 \sin C}$

$$= \frac{a^2 \sin C \sin B}{2 \sin A}$$

7. Trigonometrical Ratios of the Half Angles of a Triangle

A. *(i)* $\sin \dfrac{A}{2} = \sqrt{\dfrac{(s-b)(s-c)}{bc}}$

(ii) $\sin \dfrac{B}{2} = \sqrt{\dfrac{(s-c)(s-a)}{ca}}$

(iii) $\sin \dfrac{C}{2} = \sqrt{\dfrac{(s-a)(s-b)}{ab}}$

B. *(i)* $\cos \dfrac{A}{2} = \sqrt{\dfrac{s(s-a)}{bc}}$

(ii) $\cos \dfrac{B}{2} = \sqrt{\dfrac{s(s-b)}{ca}}$

(iii) $\cos \frac{C}{2} = \sqrt{\frac{s(s-c)}{ab}}$

C. (i) $\tan \frac{A}{2} = \sqrt{\frac{(s-b)(s-c)}{s(s-a)}}$

(ii) $\tan \frac{B}{2} = \sqrt{\frac{(s-c)(s-a)}{s(s-b)}}$

(iii) $\tan \frac{C}{2} = \sqrt{\frac{(s-a)(s-b)}{s(s-c)}}$

D. (i) $\sin A = \frac{2}{bc} \sqrt{s(s-a)(s-b)(s-c)}$

$= \frac{2\Delta}{bc}$

$\sin B = \frac{2\Delta}{ca}$

$\sin C = \frac{2\Delta}{ab}$

8. **Circumcircle of a triangle:** A circle passing through the vertices of a triangle is called the circumcircle of the triangle.

The centre of the circumcircle is called the circumcentre of the triangle and it is the point of intersection of the perpendicular bisectors of the sides of the triangle.

The radius of the circumcircle is called the circumradius of the triangle and is usually denoted by R and is given by the following formulae.

$$R = \frac{a}{2\sin A} = \frac{b}{2\sin B} = \frac{c}{2\sin C} = \frac{abc}{4\Delta}$$

9. **Incircle of a Triangle:** A circle which can be inscribed within the triangle so as to touch all the three sides is called the incircle of the triangle.

The centre of the incircle is called the incentre of the triangle and it is the point of intersection of the internal bisectors of the angles of the triangle.

The radius of the incircle is called the in-radius of the triangle and is usually denoted by r and is given by the following formulae:

$$r = \Delta/s = (s-a)\tan\frac{1}{2}A$$

$$= (s-b)\tan\frac{1}{2}B = (s-c)\tan\frac{1}{2}C.$$

$$r = \frac{a\sin\frac{1}{2}B\sin\frac{1}{2}C}{\cos\frac{1}{2}A}$$

$$= \frac{b\sin\frac{1}{2}A\sin\frac{1}{2}C}{\cos\frac{1}{2}B}$$

$$= \frac{C\sin\frac{1}{2}B\sin\frac{1}{2}A}{\cos\frac{1}{2}C}$$

$$r = 4R\sin\frac{1}{2}A\sin\frac{1}{2}B\sin\frac{1}{2}C.$$

Escribed Circles (Ex-Circle) of a triangle: The circle which touches the side BC and the other two sides AB and AC produced of a triangle ABC is called the escribed or ex-circle opposite to the angle A.

Similarly we can define the escribed circle opposite to the angles B and C.

The radii of the escribed circles opposite to the angles A, B and C are called the ex-radii and are usually denoted by r_1, r_2, r_3 respectively and are given by the following formulae–

(i) $r_1 = s\tan\frac{1}{2}A$; $r_2 = s\tan\frac{1}{2}B$;

$r_3 = s\tan\frac{1}{2}C$.

(ii) $$r_1 = \frac{a\cos\frac{1}{2}B\cos\frac{1}{2}C}{\cos\frac{1}{2}A};$$

$$r_2 = \frac{b\cos\frac{1}{2}C\cos\frac{1}{2}A}{\cos\frac{1}{2}B};$$

$$r_3 = \frac{c\cos\frac{1}{2}A\cos\frac{1}{2}B}{\cos\frac{1}{2}C}$$

(iii) $r_1 = 4R\sin\frac{1}{2}A\cos\frac{1}{2}B\cos\frac{1}{2}C$,

$$r_2 = 4R \cos \frac{1}{2} A \sin \frac{1}{2} B \cos \frac{1}{2} C,$$

$$r_3 = 4R \cos \frac{1}{2} A \cos \frac{1}{2} B \sin \frac{1}{2} C.$$

Cyclic Quadrilateral: A quadrilateral ABCD is said to be a cyclic quadrilateral if there a circle passing through all its four vertices A, B, C and D.

(i) Area of a cyclic quadrilateral ABCD

$$= \sqrt{(s-a)(s-b)(s-c)(s-d)}$$

where AB = a, BC = b, CD = c, DA = d and $2s = a + b + c + d$

(ii) Circumradius of a cyclic quadrilateral

$$R = \frac{1}{4} \sqrt{\frac{(ab+cd)(ac+bd)(ad+bc)}{(s-a)(s-b)(s-c)(s-d)}}$$

Regular Polygon: A polygon is said to be a regular polygon if all its sides are equal and its angles are equal.

(i) *Circumscribed Circle:* The circle passing through all the vertices of a regular polygon is called its circumscribed circle.

If a is the length of each side of a regular polygon of n sides, then the radius R of the circumscribed circle is given by

$$R = \frac{a}{2\sin(\pi/n)} = \frac{a}{2}\operatorname{cosec}(\pi/n)$$

(ii) *Inscribed Circle:* The circle which can be inscribed within the regular polygon so as to touch all its sides is called its inscribed circle.

The radius r of inscribed circle of a regular polygon of n sides, each of length a is given by

$$r = \frac{a}{2}\cot\left(\frac{\pi}{n}\right)$$

(iii) *Area of a regular polygon:* The area, Δ of a regular polygon of n sides, each of length a is given by

$$\Delta = \frac{1}{4}na^2\cot(\pi/n)$$

$$= \frac{1}{2}nR^2\sin(2\pi/n)$$

$$= nr^2\tan(\pi/n)$$

11

CARTESIAN SYSTEM OF RECTANGULAR COORDINATES

oordinate geometry is that branch of geometry which two numbers called co-ordinates, are sed to indicate position of a point in a plane and hich makes the use of algebraic methods in the udy of geometric figures.

istance Formula: Distance between two oints P and Q whose coordinates are (x_1, y_1) and (x_2, y_2) is

$$PQ = \sqrt{(x_2 - x_1)^2 + (y_2 - y_1)^2}$$

istance of a point $p\ (x_1, y_1)$ from Origin: The istance of a point P (x_1, y_1) from the origin (0, 0) s

$$\sqrt{(x_1 - 0)^2 + (y_1 - 0)^2} = \sqrt{x_1^{\,2} + y_1^{\,2}}$$

Remember:

(*i*) Distance between two given points

$= \sqrt{(\text{Difference of Abscissa})^2 + (\text{Difference of Ordinates})^2}$

(*ii*) We take only the positive square root as we are usually interested only in the magnitude of the line PQ.

Application of Distance formula to Geometry:

(A) When three points are given

Case I: To show that the given three points A, B, C form an isosceles triangle

Working Rule

Step I: Find the length of the sides AB, BC, CA with the help of distance formula.

Step II: Show that two of the three sides AB, BC, CA are equal. Then the triangle is Isosceles triangle.

Case II: To show the given three points A, B and C form a right angled triangle.

Working Rule:

Step I: Find the sides AB, BC, CA.

Step II: Show that the square of one side is equal to the sum of the squares of the other two. Then ΔABC is a Right angled triangle.

Case III: To show that the given three points A, B, C form an equilateral triangle.

Working Rule:

Step I: Find the sides AB, BC, CA.

Step II: Show that AB = BC = CA. Then ΔABC is an Equilateral triangle.

Case IV: To show that given three points A, B, C form a right-angled isosceles triangle.

Working Rule:

Step I: Find the sides AB, BC, CA.

Step II: Show that two of the three sides are equal.

Step III: Show that the square of one side is equal to the sum of the squares of the other two sides (Pythagorus Theorem)

The ΔABC is a Right-angled Isosceles triangle.

Case V: To show that the three poins A, B, C are collinear.

Step I: Find AB, BC, CA.

Step II: Show that the sum of the distances between two pairs of points is equal to the distance between the third pair. Or, show that the area of the triangle formed by these three poir is zero.

(B) When four points are given

Case VI: To show that the four points B, C, D form a parallelogram.

Working Rule:

Step I: Find AB, BC, CD, DA.

Step II: Show that the opposite sid are equal, for example, AB = DC aı BC = DA, then the four points A, B, D form a parallelogram. Or, to sho that the mid-point of the diagonal A is equal to mid point of the diagonal BI

Case VII: To show that the four points for a rectangle.

Step I: Find AB, BC, CD, DA.

Step II: Show that the opposit diagonals are equal. Or, to show tha ΔABC is a right angled triangle i.e., $AB^2 + BC^2 = CA^2$

Hence, the quadrilateral is a Rectangl

Case VIII: To show that the four points A B, C, D form a rhombus.

Working Rule:

Step I: Find the sides AB, BC, CD an DA.

Step II: Show that all the four sides are equal *i.e.*, AB = BC = CD = DA.

Hence, the quadrilateral ABCD is a rhombus.

Case IX: To show that the four points A, B, C, D form a square.

Working Rule:

Step I: Find the sides AB, BC, CD, DA.

Step II: Show that all the sides are equal.

Step III: Also, show that the diagonals are equal. Or, to show that ΔABC is a right angled triangle. Hence, the quadrilateral is a square.

Section Formulae:

(*i*) The coordinates of the point P (x, y), dividing the line joining the points A (x_1, y_1) and B (x_2, y_2) in the ratio $m : n$ (Internally and Externally).

(*a*) If P divides AB internally, then

A(x_1, y_1) — m — P(x, y) — n — B(x_2, y_2)

$$x = \frac{mx_2 + nx_1}{m+n}$$

$$y = \frac{my_2 + ny_1}{m+n}$$

(b) If P divides AB externally, then

m : n
A(x_1, y_1) B(x_2, y_2) P (x, y)

$$x = \frac{mx_2 - nx_1}{m-n}$$

$$y = \frac{my_2 - ny_1}{m-n}$$

(ii) The coordinates of the mid-point of line joining A (x_1, y_1) and B (x_2, y_2) are

$$\left(\frac{x_1 + x_2}{2}, \frac{y_1 + y_2}{2}\right)$$

Remember:

(i) The line joining any vertex of a triangle to the mid-point of the opposite side is called a median of the triangle. Therefore, a triangle has three medians.

(ii) The point of intersection of the medians of a triangle is called the centroid of a triangle.

(iii) The centroid of a triangle divides each median in the ratio 2 : 1.

The coordinates of the centroid G of a triangle whose vertices are (x_1, y_1), (x_2, y_2), and (x_3, y_3) is $\left[\dfrac{x_1 + x_2 + x_3}{3}, \dfrac{y_1 + y_2 + y_3}{3}\right]$

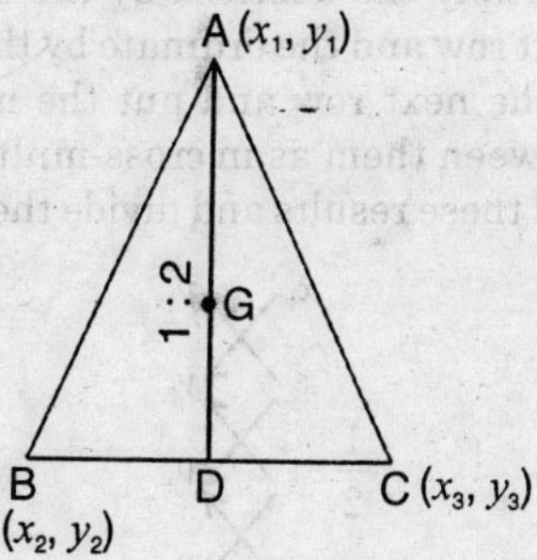

Trapezium: The area of a Trapezium

$= \dfrac{1}{2}$ (sum of parallel sides) × (perpendicular distance between them)

Area of a triangle: The area of a triangle whose vertices are (x_1, y_1), (x_2, y_2) and (x_3, y_3) is given by

$$\Delta = \frac{1}{2} [(x_1 y_2 - x_2 y_1) + (x_2 y_3 - x_3 y_2) + (x_3 y_1 - x_1 y_3)]$$

Remember: An easier method of writing down the area of a triangle quickly is as follows:

(i) Write down the coordinates of the vertices as shown repeating at the end the coordinates at the top.

(ii) Multiply the abscissa by the ordinate of next row and the ordinate by the abscissa of the next row and put the minus sign between them as in cross-multiplication; add these results and divide the sum by 2.

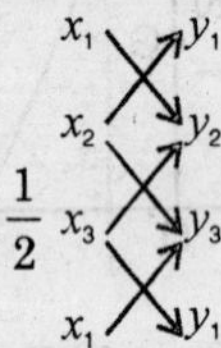

Thus $\begin{matrix} x_1 & y_1 \\ x_2 & y_2 \end{matrix} = x_1 y_2 - x_2 y_1 + \ldots$

Condition for collinearity of three points: The condition that the three points lie in a straight line is

$$x_1 y_2 - x_2 y_1 + x_2 y_3 - x_3 y_2 + x_3 y_1 - x_1 y_3 = 0$$

[If three points are collinear, the area of the triangle formed by them is zero].

Area of the quadrilateral: The area of the quadrilateral whose vertices (x_1, y_1) (x_2, y_2) (x_3, y_3) and (x_4, y_4) is given by $[x_1y_2 - x_2y_1 + x_2y_3 - x_3y_2 + x_3y_4 - x_4y_3 + x_4y_1 - x_1y_4]$

Remember: The rule to write the area of a quadrilateral is the same as the rule for triangle.

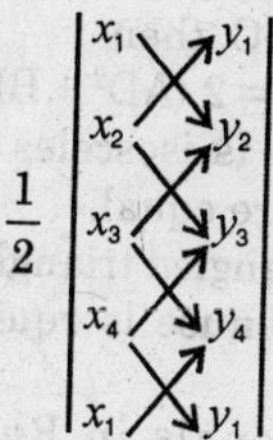

Locus and its Equation: *Locus:* The path described by a point which moves under a given condition or conditions is called its locus.

Equation of a Locus: The equation of a locus is an equation in x and y which is satisfied by the coordinates of any and every point on the locus and by the coordinates of no other point.

Method to find the equation of the locus of a moving point: To find the locus of a point proceed as follows:

1. Let P (x, y) be any point on the locus.
2. Write down the given condition under which the point P moves.
3. Express the said condition in terms of x and y and simplify the results.
4. The simplified result so obtained is the required equation of the locus.
5. In a triangle ABC, if AD is the median drawn to BC, then
   $AB^2 + AC^2 = 2\,(AD^2 + BD^2)$
6. A triangle is isosceles if any two of its medians are equal.
7. In a right angled triangle, the midpoint of the hypotenuse is equidistant from the vertices.
8. The diagonals in Rhombus, Square, Rectangle and Parallelogram bisect each other.
9. The figure obtained by joining the middle points of a quadrilateral in order is parallelogram.

10. If (x, y) are the coordinates of a point P w.r.t. the axes OX and OY and (x', y') its coordinates w.r.t. the axes OX′.
OY′ obtained by rotating the original axes OX, OY through an angle θ in anticlockwise direction, keeping the same origin, then

$$x = x' \cos \theta - y' \sin \theta$$

$$y = x' \sin \theta + y' \cos \theta$$

Thus, to find the equation of the curve referred to the turned axes, through angle θ, replace
x by $x' \cos \theta - y' \sin \theta$
and$y$ by $x' \sin \theta + y' \cos \theta$
The relations connecting (x, y) and (x', y') can be written in the following scheme

	x'	y'
x	$\cos \theta$	$-\sin \theta$
y	$\sin \theta$	$\cos \theta$

11. To remove the term of xy in the equation, $ax^2 + 2hxy + by^2 = 0$, the angle θ through which the axes must be turned (rotated) is given by

$$\theta = \frac{1}{2} \tan^{-1} \{2h/(a - b)\}.$$

12

THE STRAIGHT LINE

The general form of the equation of a straight line is $ax + by + c = 0$ (the first degree equation in x and y) where a, b and c are real constants not all equal to zero simultaneously.

The line parallel to axes:

(i) The equation of a straight line $||^l$ to x-axis and at a distance b from it is given by $y = b$.

(ii) The equation of the straight line $||^l$ to y-axis and at a distance a from it is given by $x = a$.

(iii) The equation of the straight line $||^l$ to x-axis and y-axis respectively and passing through the pt. (a, b) are $y = b$, $x = a$.

(iv) Equation of x-axis is $y = 0$ and equation of y-axis is $x = 0$.

Slope intercept form: The equation of a straight line which cuts off an intercept c on y-axis and

makes an angle θ with the positive direction of x-axis is given by $y = mx + c$, where $m = \tan\theta$.
If θ is acute, then tan θ *i.e.,* m is positive.
If θ is obtuse, then tan θ *i.e.,* m is negative.
The number $m = \tan\theta$ is called the slope or gradient of line obviously $y = mx$ is the equation of the line through the origin.

Note: *(i)* If a line is parallel to x-axis $\theta = 0$ and its slope $m = \tan\theta = 0$

(ii) If a line parallel to y-axis it is perpendicular to x-axis so that $\theta = 90°$ and its slope $m = \tan 90° = \infty$

Condition of parallelism and perpendicularity of lines

Case I: Slopes of parallel lines are equal.
i.e., $m_1 = m_2$

Case II: The product of the slope of two perpendicular lines is –1.
i.e., $m_1 m_2 = -1$

Slope point form: The equation of a straight line passing through the point A (x_1, y_1) and having slope m is given by

$$y - y_1 = m(x - x_1)$$

This equation can also be written as

$$\frac{x-x_1}{\cos\theta}=\frac{y-y_1}{\sin\theta}=r\text{ (say)}\qquad [\because m=\tan\theta]$$

This form of the line is called parametric form or symmetrical form of the line.
The coordinates of any point P on this line are taken as $(x_1 + r\cos\theta, y_1 + r\sin\theta)$

$\therefore$ Distance of a point P from A (x_1, y_1)

$$=\sqrt{\left[(x_1+r\cos\theta-x_1)^2+(y_1+r\sin\theta-y_1)^2\right]}=r$$

Thus, if the equation of a line through point A (x_1, y_1) is written in the above form, in which the coefficients of x and y are equal to 1 and the sum of squares of the terms in the denominators is equal to 1, then the coordinates of a point P on this line at a distance r from A (x_1, y_1) are

$$(x_1 + r_1\cos\theta, y_1 + r\sin\theta)$$

Two Point Form: The equation of a straight line passing through two points A (x_1, y_1) and B (x_2, y_2) is given by

$$y-y_1=\frac{y_2-y_1}{x_2-x_1}(x-x_1).$$

$$\text{Here, slope } (m)=\frac{y_2-y_1}{x_2-x_1}=\frac{\text{difference of ordinates}}{\text{difference of abscissae}}$$

Intercept form: The equation of a straight line making intercepts a and b on the axes of x and y respectively is given by

$$\frac{x}{a}+\frac{y}{b}=1$$

Normal (or perpendicular) form: The equation of a straight line on which the perpendicular from origin is of length p and makes an angle α with the positive direction of x-axis is given by

$$x\cos\alpha + y\sin\alpha = p$$

Remember:

(i) If $\alpha = 0°$, the equation becomes
$x\cos 0° + y\sin 0° = p$ or, $x = p$
which is a line parallel to y-axis

(ii) If $\alpha = \pi/2$, the equation becomes
$x\cos \pi/2 + y\sin \pi/2 = p$
or, $y = p$
Which is a line parallel to x-axis.

(iii) If $\alpha = 0°$, $p = 0$, the equation becomes
$x\cos 0° + y\sin 0° = 0$
or, $x = 0$
which is the equation of y-axis

(iv) If $\alpha = \pi/2$, $p = 0$, the equation becomes
$x\cos \pi/2 + y\sin \pi/2 = 0$

or, $y = 0$

which is the equation of x-axis.

General Equation of a straight line: The equation of the first degree in x and y represent a straight line *i.e.,*

$$Ax + By + C = 0$$

where A, B, C $\in$ R and A, B are not zero simultaneously.

Reduction of the General Equation of the straight line to standard forms:

(*a*) Reduction of general equation of a straight line to the slope intercept form is

$$y = -\frac{A}{B}x - \frac{C}{B}$$

$$\text{Here, slope} = -\frac{A}{B} = -\frac{\text{coeff. of } x}{\text{coeff. of } y}$$

(*b*) Reduction of general equation of a straight line to the intercept form is

$$\frac{x}{\frac{-C}{A}} + \frac{y}{\frac{-C}{B}} = 1$$

(*c*) Reduction of the general equation of a straight line to the normal (or perpendicular) form $p^2 (A^2 + B^2) = C^2$

$$\therefore \quad p = \pm \frac{C}{\sqrt{A^2+B^2}}$$

But p is always positive.

Remember:

To reduce the general equation of a line to the normal form:

(i) Transpose the constant term to R.H.S. and make it positive. If the constant term **is** negative after transposition multiply both sides by (–1) to make it positive.

(ii) Dividing throughout by

$$\sqrt{A^2+B^2} = \sqrt{(\text{coeff. of } x)^2+(\text{coeff. of } y)^2}$$

Angle between two straight lines: The angle between the lines

$$y = m_1 x + c_1 \text{ and } y = m_2 x + c_2 \text{ is}$$

$$\tan\theta = \frac{m_1 - m_2}{1+m_1 m_2}$$

Note: If $\tan\theta$ is +ve, θ is acute.

If $\tan\theta$ is –ve, θ is obtuse.

$\therefore$ The acute angle between the two lines is

given by $\tan\theta = \left|\dfrac{m_1 - m_2}{1 - m_1 m_2}\right|$

(a) **Condition of parallelism:**

If lines are parallel; $\theta = 0°$ or $180°$

$\therefore \quad \tan\theta = 0$ or $\tan 180°$

$\therefore \quad m_1 = m_2$

(b) **Condition of perpendicularity:**

If lines are perpendicular, $\theta = 90°$

$\therefore \quad \tan\theta = \tan 90° = \infty$

$\therefore \quad m_1 m_2 = -1$

Equation of line parallel to $Ax + By + C = 0$:

The equation of the straight line which is parallel to the line $Ax + By + C = 0$ is given by $Ax + By + \lambda = 0$. Thus, to write the equation of any line parallel to a given line, do not change the coefficients of x and y and change to constant term only.

Equation of line perpendicular to $Ax + By + C = 0$:

The equation of a straight line which is perpendicular to the line $Ax + By + C = 0$ is given by $Bx - Ay + \lambda = 0$. Thus, to write the equation of any line perpendicular to given line interchange the coefficient of x and y in the given equation and change the sign of one of those coefficient and also change the constant term.

Point of intersection of two given lines: Let the two given lines be $a_1x + b_1 y + c_1 = 0$ and $a_2x + b_2y + c_2 = 0$. Solving these equations, the point of intersection of these two lines is given by

$$\left(\frac{b_1c_2 - b_2c_1}{a_1b_2 - a_2b_1}, \frac{c_1a_2 - c_2a_1}{a_1b_2 - a_2b_1}\right)$$

Concurrent lines: Three or more straight lines are said to be concurrent lines if they meet in a point.

Method to prove the three lines to be concurrent:

(i) Find the point of intersection of any two lines and if this point satisfy the equation of the third line, then the three lines are concurrent, and this point is the point of intersection of the three lines.

(ii) The three lines $a_1x + b_1y + c_1 = 0$, $a_2x + b_2y + c_2 = 0$ and $a_3x + b_3y + c_3 = 0$ are concurrent if

$$\begin{vmatrix} a_1 & b_1 & c_1 \\ a_2 & b_2 & c_2 \\ a_3 & b_3 & c_3 \end{vmatrix} = 0.$$

(iii) Let P = 0, Q = 0 and R = 0 be three given lines. Then these lines are concurrent if we can find three constants l, m, n not all zero (*i.e.*, at least one of these is not equal to zero) such that $lP + mQ + nR = 0$ takes the form $0.x + 0.y + 0 = 0$.

Any line through the point of intersection of two lines: Let P = 0 and Q = 0 be the two given lines. Then the equation of any line passing through the point of intersection of these two lines is $P + \lambda Q = 0$, where λ is any real number.

The value of λ will be obtained by the condition given in the question.

The length of perpendicular from a given point (x_1, y_1) to a line $ax + by + c = 0$ is

$$\frac{ax_1 + by_1 + c}{\sqrt{(a^2 + b^2)}}$$

Distance between parallel lines:

Let the two parallel lines be

$ax + by + c_1 = 0$ and $ax + by + c_2 = 0$.

Now, to find the distance between these two parallel lines, we proceed as follows:

(A) First Method: Find a point on any of the line. For this put $x = 0$ and find y, put $y = 0$

and find x. Then the distance between the lines is equal to the length of perpendicular from this point on one line to the other line.

(B) IInd Method: Let P (h,k) be a point on one line, say $ax + by + c_1 = 0$,

$$\therefore \quad ah + bk = -c_1$$

Then, distance between lines

= Length of perpendicular from P (h, k) to the other line $ax + by + c_2 = 0$

$$= \left|\frac{ah + bk + c_2}{\sqrt{(a^2 + b^2)}}\right| = \left|\frac{c_2 - c_1}{\sqrt{(a^2 + b^2)}}\right|$$

(C) IIIrd Method: Find the distance of each line from the origin. If these distances are p_1 and p_2, then $p_1 = \dfrac{c_1}{\sqrt{(a^2 + b^2)}}$ and $p_2 = \dfrac{c_2}{\sqrt{(a^2 + b^2)}}$

and retain their signs.

Then, the perpendicular distance between

lines $= |p_2 - p_1| = \left|\dfrac{c_2 - c_1}{\sqrt{a^2 + b^2}}\right|$

Equation of the bisectors of the angles between two lines:

Let the two given lines be

$a_1x + b_1y + c_1 = 0$ and $a_2 x + b_2 y + c_2 = 0$

Then, the equation of the bisectors of the angles between two lines are

$$\frac{a_1x + b_1y + c_1}{\sqrt{(a_1^{\;2} + b_1^{\;2})}} = \pm \frac{a_2x + b_2y + c_2}{\sqrt{(a_2^2 + b_2^2)}}$$

Remember that, every point of the bisectors of angles is equidistant from the given lines.

FAMILY OF LINES

We know that the eqn. $y = mx + 4$ represents a straight line having slope m and making an intercept 4 on y-axis.

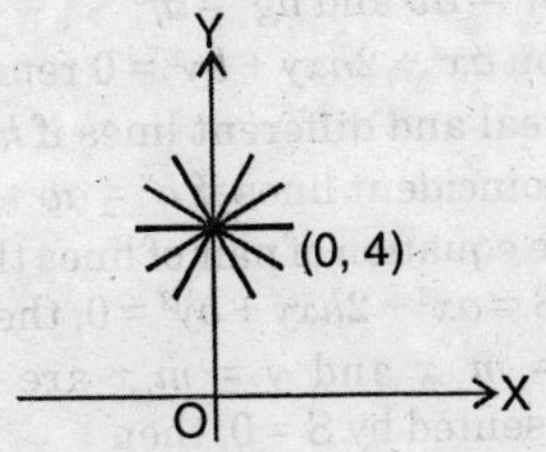

For different value of m represents different straight lines each making an intercept 4 on y-axis. All these lines (as shown in figure) taken together constitute a family of lines having the common property that each makes an intercept of 4 units on y-axis.

The arbitrary constant m, which is same for one line but is different for different lines, is called the parameter of the family.

Let $\quad S = ax^2 + 2hxy + by^2 + 2gx + 2fy + c = 0$

and $\quad \Delta = abc + 2fgh - af^2 - bg^2 - ch^2$.

Then, S = 0 represents.

(i) a pair of straight lines, if $\Delta = 0$ and $h^2 \geq ab$

(ii) a pair of intersecting lines, if $\Delta = 0$ and $h^2 > ab$

(iii) a pair of parallel lines, if $\Delta = 0$ and $h^2 = ab$ or if $h^2 = ab$ and $bg^2 = af^2$

The equation $ax^2 + 2hxy + by^2 = 0$ represents

(i) two real and different lines if $h^2 > ab$

(iii) two coincident lines if $h^2 = ab$

Consider the equation of pair of lines through the origin, *i.e.*, $S = ax^2 + 2hxy + by^2 = 0$, then

(i) if $y = m_1 x$ and $y = m_2 x$ are two lines represented by S = 0, then

$m_1 + m_2 = -2h/b$ and $m_1 m_2 = a/b$

(ii) The angle θ between the lines represented by S = 0, is given by $\theta = \tan^{-1}\left|\dfrac{2\sqrt{(h^2 - ab)}}{a+b}\right|$

(iii) The lines given by S = 0 are parallel if $h^2 = ab$.

(iv) The line given by S = 0 are perpendicular if $a + b = 0$, *i.e.*, if coeff. of x^2 + coeff. of $y^2 = 0$

(v) The equation of the pair of bisectors of the angles between the lines S = 0 is

$$\frac{x^2 - y^2}{a-b} = \frac{xy}{h}.$$

14

CIRCLES AND FAMILY OF CIRCLES

Circle: A circle is a locus of a point which moves in a plane so that it is always at a constant distance from a fixed point in the plane. The fixed point is called the centre and the constant distance is called the radius of the circle.

Equation of a Circle: The equation of a circle whose centre is C (h, k) and radius r is

$$(x-h)^2 + (y-k)^2 = r^2$$

Particular Case: If a centre of the circle is origin (0, 0), then the above equation reduces to

$$(x-0)^2 + (y-0)^2 = r^2$$

or
$$x^2 + y^2 = r^2$$

General Equation of a Circle: $x^2 + y^2 + 2gx + 2fy + c = 0$ is the equation of a circle whose centre is $(-g, -f)$ and radius is $\sqrt{g^2 + f^2 - c}$

Note:

(A) To find the centre and radius of circle

(i) make the coefficient of x^2 and y^2 each equal to 1 by dividing throughout (if necessary) by the coefficients of x^2 and y^2.

(ii) The coordinates of the centre are

$$\left(-\frac{1}{2}\text{coeff. of } x, \frac{-1}{2}\text{coeff. of } y\right)$$

(iii) Radius

$$= \sqrt{\left(-\frac{1}{2}\text{coeff. of } x\right)^2 + \left(-\frac{1}{2}\text{coeff. of } y\right)^2 - \text{constant term}}$$

(B) Nature of the circle

(i) If $g^2 + f^2 - c = 0$, the radius is zero and the circle becomes a point, which is known as a point circle.

(ii) If $g^2 + f^2 - c > 0$, the radius is real and hence, the cicle is a real circle.

(iii) If $g^2 + f^2 - c < 0$, the radius is imaginary and therefore, the circle is an imaginary circle and is called a virtual circle.

Condition of a Circle: Any equation satisfying the following conditions must be a circle.

(i) The coeff. of x^2 = coeff. of y^2

(ii) It contains no term involving the product xy.

(iii) It is an equation of second degree in x and y.

Diameter form of a Circle: The equation of a circle whose end diameter are (x_1, y_1) and (x_2, y_2) is given by $(x - x_1)(x - x_2) + (y - y_1)(y - y_2) = 0$

Intersection of a line and a circle:

(i) The line $y = mx + c$ will intersect the circle $x^2 + y^2 = r^2$ in two distinct and real points if $c^2 < r^2 (1 + m^2)$

(ii) The line $y = mx + c$ will intersect the circle $x^2 + y^2 = r^2$ at one point if $c^2 = r^2 (1 + m^2)$

(iii) The line $y = mx + c$ will intersect the circle $x^2 + y^2 = r^2$ at two imaginary points if $c^2 > r^2 (1 + m^2)$

Tangent to a Circle:

Let P be any point on a curve and Q be any other point on the curve, join PQ and that PQ is called a chord of the Circle.

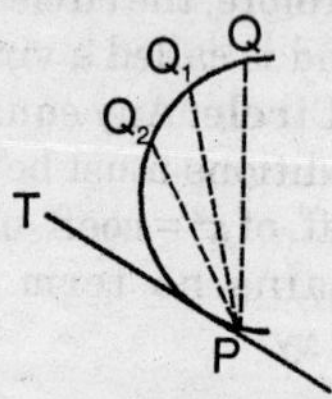

Let the point Q moves along the curve towards P taking the position Q_1, Q_2 ... and finally coincide with P such that the chord PQ takes the position PT. The limiting position PT of the line PQ as Q tends to P is called the tangent to the curve at P.

Equation of a tangent

(i) The equation of the tangent at point (x_1, y_1) on the circle is $x x_1 + y y_1 = r^2$

(ii) Equation of the tangent at the point (x_1, y_1) on the circle $x^2 + y^2 + 2gx + 2fy + c = 0$ is $xx_1 + yy_1 + g(x + x_1) + f(y + y_1) + c = 0$.

Working Rule:

(i) Change x^2 into xx_1 and y^2 into yy_1

(ii) Change x into $\frac{x + x_1}{2}$ and y to $\frac{y + y_1}{2}$ and keep constant term same.

Normal: The normal at a point of a curve is the straight line through the point and perpendicular to the tangent at the point.

Equation of a Normal

(i) Equation of a normal at the point (x_1, y_1) to the circle $x^2 + y^2 = r^2$ is

$$xy_1 - yx_1 = 0$$

Note: (a) The normal at any point $(x_1 y_1)$ to the circle $x^2 + y^2 + 2gx + 2fy + c = 0$ is

$$x(y_1 + f) - y(x_1 + g) + (gy_1 - fx_1) = 0$$

A Circle and a line: A line $y = mx + c$ will be a tangent to the circle $x^2 + y^2 = a^2$ is $c = \pm a\sqrt{1+m^2}$

In other words, $y = mx \pm a\sqrt{1+m^2}$ is equation of the tangent to the circle $x^2 + y^2 = a^2$.

This is called the equation of the tangent to the circle in slope form.

Parametric equation of a circle:

(i) Parametric equations of a circle $x^2 + y^2 = a^2$ are $x = a \cos \theta$ and $y = a \sin \theta$.

(ii) The parametric equations of a circle $(x - h)^2 + (y - k)^2 = a^2$ are $x = h + a \cos \theta$ and $y = k + a \sin \theta$ where θ is a parameter.

Radical axis of two circles and its equation:

The radical axis of two non-concentric circles is the locus of a point which moves such that the lengths of the tangents drawn from it to the two circles are equal.

The equation of radical axis of the circle is

$x^2 + y^2 + 2g_1x + 2f_1y + c_1 = 0$ and

$x^2 + y^2 + 2g_2x + 2f_2y + c_2 = 0$

$2(g_1 - g_2)x + 2(f_1 - f_2)y + (c_1 - c_2) = 0$

Rule to find the equation of radical axis:

(i) Subtract the equations of two circles, after making coefficients of x^2 and y^2 unity in both the equations.

(ii) If the two circles intersects, then radical axis is the common chord of the two circles.

(iii) If the two circles touch each other, then radical axis is the common tangent.

Radical Centre of three circle: The point of intersection of radical axes of three circles taken in pairs is called the radical centre of the three circles.

Family of Circles: A collection of circles is called a family of circles.

The general equation of a circle is

$x^2 + y^2 + 2gx + 2fy + c = 0$ where g, f, c are three arbitrary constants (called parameters). We also know that three constants g, f, c can have fixed value only when three independent conditions satisfied by a circle are given.

15

CONIC SECTION

A Conic Section or conic is the locus of a point P which moves so that its distance from a fixed point S is in a constant ratio to its perpendicular distance from a fixed straight line, all being in the same plane.

The fixed point is called the focus and is generally denoted by S. The fixed straight line is called the directrix and the constant ratio is called the ecentricity and is denoted by the letter e.

The straight line passing through the focus and perpendicular to the directrix is called the axis of the conic, and the point where the axis meets the conic is called the vertex of the conic.

The conic is called a parabola if $e = 1$

The conic is called a ellipse if $e < 1$

Y
M
P(x, y)
X
Z A
S (a,0)

The conic is called a hyperbola if $e > 1$.

Parabola: A parabola is the locus of a point which moves in a plane so that its distance from a fixed point is equal to its distance from a fixed straight line.

The fixed point is called the focus and fixed straight line is called the directrix of the parabola.

Equation of the parabola: It is $y^2 = 4ax$

Note:

(i) The coordinates of the vertex A are (0,0).

(ii) The coordinates of the focus S are $(a, 0)$.

(iii) The equation of the axis of the parabola is $y = 0$

(iv) The equation of the directrix is $x = -a$.

Definitions:

1. **Focal distance:** The distance of a point on the parabola from the focus is called the focal distance of the point.
2. **Focal chord:** A chord of the parabola which passes through the focus is called a focal chord.
3. **Latus rectum:** The latus rectum of a parabola is a chord passing through the focus and perpendicular to the axis.

Four standard forms of the parabola

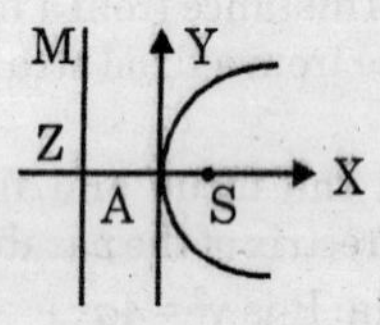

(*i*) $y^2 = 4ax$

[Right handed parabola]

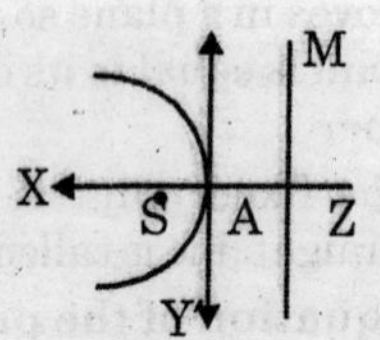

(*ii*) $y^2 - 4ax$

[Left handed parabola]

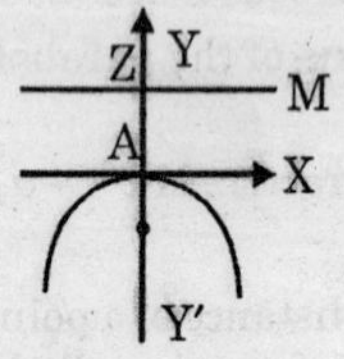

(*i*) $x^2 - -4ay$

[Down parabola]

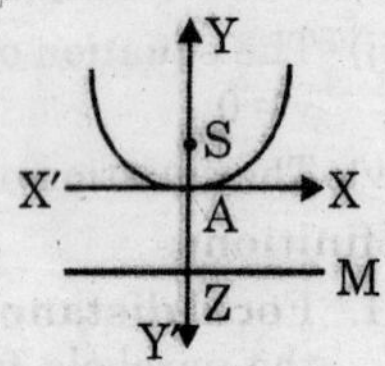

(*i*) $x^2 = 4ay$

[Upward parabola]

Important facts relating to parabola

Form	$y^2 = 4ax$	$y^2 = -4ax$	$x^2 = 4ay$	$x^2 = -4ay$
Focus	$(a, 0)$	$(-a, 0)$	$(0, a)$	$(0, -a)$
Directrix	$x = -a$	$x = a$	$y = -a$	$y = a$
Axis	$y = a$	$y = 0$	$x = 0$	$x = 0$
Vertex	$(0,0)$	$(0,0)$	$(0,0)$	$(0,0)$
Tangent at vertex	$x = 0$	$x = 0$	$y = 0$	$y = 0$
Rectum	$4a$	$4a$	$4a$	$4a$

Ellipse: An ellipse is the locus of a point which moves in a plane so that its distance from a fixed point bears a constant ratio, less than one, to its distance from a fixed straight line.

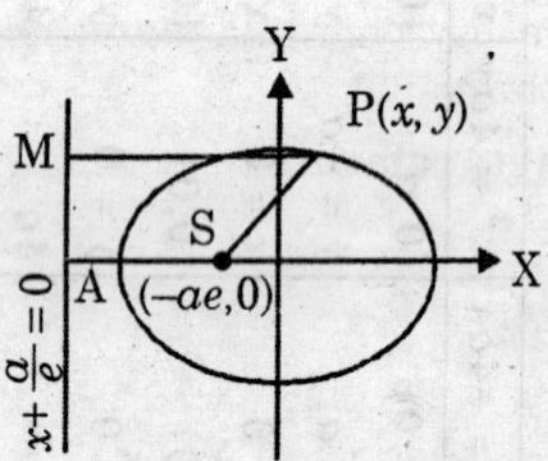

If P is any point on the ellipse whose focus is S, and ecentricity is e and PM ⊥ on the directrix, then

$$\frac{SP}{PM} = e \text{ or } SP = e\,PM (e < 1)$$

Equation of the ellipse:

$$\frac{x^2}{a^2} + \frac{y^2}{b^2} = 1$$

Definitions:

(*i*) The lines AA′ and BB′ are called the major and minor axis. Both together are called principal axis.

(*ii*) The points A, A′, the extremities of the major axis are called the vertices of the ellipse.

(*iii*) C is centre of the ellipse. The centre C bisects every chord of the ellipse which passes through it.

Important facts relating to an ellipse:

Ellipse: $\frac{x^2}{a^2}+\frac{y^2}{b^2}=1$ **Ellipse:** $\frac{x^2}{b^2}+\frac{y^2}{a^2}=1$

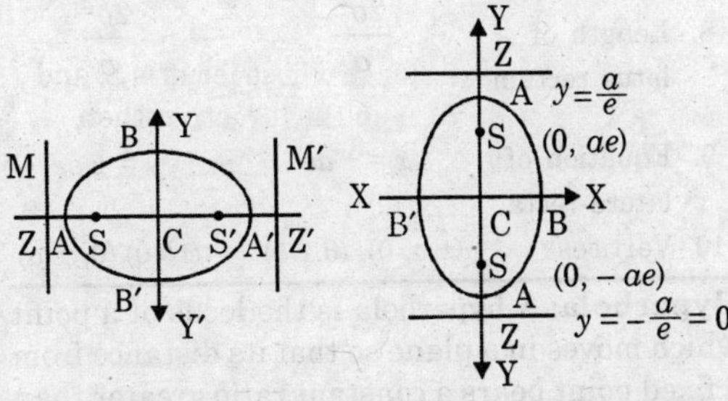

1. Centre	(0, 0)	(0, 0)
2. Foci	$(\pm ae, 0)$	$(0 \pm ae)$
3. Length of major axis	$2a$	$2a$

4. Length of minor axis	$2b$	$2b$
5. Equation of major axis	$y = 0$	$x = 0$
6. Equation of minor axis	$x = 0$	$y = 0$
7. Equation of directrices	$x = \pm\frac{a}{e}$	$y = \pm\frac{a}{e}$
8. Length of latus rectum	$\frac{2b^2}{a}$	$\frac{2b^2}{a}$
9. Equation of latera-recta	$x = \pm\, ae$	$y = \pm\, ae$
10. Vertices	$(\pm\, a, 0), (0, \pm\, b)$	$(\pm b, 0), (0, \pm\, a)$

Hyperbola: A hyperbola is the locus of a point which moves in a plane so that its distance from a fixed point bears a constant ratio greater than one, to its distance, from a fixed straight line.

The fixed point is called the focus and is generally denoted by S. The fixed straight line is called the directrix of the hyperbola and the constant ratio is usually denoted by e, is called the ecentricity of the hyperbola.

Equation of hyperbola in standard form: is

$$\frac{x^2}{a^2}-\frac{y^2}{b^2}=1$$

Note:

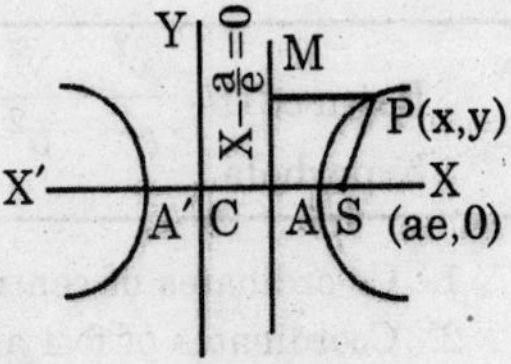

The y-axis *i.e.*, $x = 0$ meets the hyperbola

Y, M, P(x,y), X′, X, A′, C, A, S, (ae,0), $X-\frac{a}{e}=0$

$$\frac{x^2}{a^2}-\frac{y^2}{b^2}=1$$

where $y^2 = -b^2$ or $y = \pm ib$

∴ The points of intersection are imaginary.

Note:

(*i*) **Vertices:** The point A and A' are called vertices of the hyperbola.

(*ii*) **Transverse axis:** AA' is called the transverse axis of the hyperbola.

(*iii*) **Conjugate axis:** Y is called conjugate of the hyperbola.

(*iv*) **Principal axis:** The transverse and conjugate axes together are called the principal axes of the hyperbola.

(*v*) **Centre:** C is called the centre of the hyperbola. It is the point of intersection of the transverse and conjugate axes. It bisects every chord of the hyperbola that passes through it.

Important facts relating to hyperbola

Form of the hyperbola	$\frac{x^2}{a^2}-\frac{y^2}{b^2}=1$	$\frac{y^2}{a^2}-\frac{x^2}{b^2}=1$
1. Co-ordinates of centre are	(0, 0)	(0, 0)
2. Coordinates of foci are	($\pm$ *ae*, 0)	(0, $\pm$ *ae*)
3. Coordinates or vertices are	($\pm$ *a*, 0)	(0, $\pm$ *a*)
4. Equation of transverse axis	$y = 0$	$x = 0$
5. Equation of conjugate axis	$x = 0$	$y = 0$
6. Equation of the directrices	$x=\pm\frac{a}{e}$	$y=\pm\frac{a}{e}$

7. Ecentricity is given by $b^2 = a^2 (e^2 - 1)$ or

$$e^2=\frac{a^2+b^2}{a^2}$$

8. Length of latus rectum $=\frac{2b^2}{a}$
9. Length of transverse axis = $2a$
10. Length of conjugate axis = $2b$

11. For a rectangular hyperbola $b = a$ and $e = \sqrt{2}$
12. Equation of hyperbola with centre, (h,k) is

$$\frac{(x-h)^2}{a^2} = \frac{(y-h)^2}{b^2} = 1$$

13. The coordinates of the ends of latera recta are:

$$\left(ae, \frac{b^2}{a}\right); \left(ae, \frac{-b^2}{a}\right); \left(-ae, \frac{b^2}{a^2}\right); \left(-ae, \frac{-b^2}{a^2}\right)$$

14. The equation of latera recta are $x = \pm\, ae$
15. Focal distances of any point $P(x, y)$ on the hyperbola are $ex \pm a$

Condition of Tangency of the line $y = mx + c$

(*a*) **Parabola:** Condition that line $y = mx + c$ is tangent to the parabola $y^2 = 4ax$ is $c = \dfrac{a}{m}$.

and the point of contact is $\left(\dfrac{a}{m^2}, \dfrac{2a}{m}\right)$.

Parametric coordinates: Any point on the parabola $y^2 = 4ax$ is $(at^2, 2at)$ and is generally referred to as the point '*t*'.

(b) Ellipse: Condition that line $y = mx + c$ is tangent to the ellipse $\frac{x^2}{a^2}+\frac{y^2}{b^2}=1$ is

$$c = \pm\sqrt{a^2m^2+b^2}$$

Point of contact: The point of contact of this tangent with the ellipse is

$$\left(-\frac{a^2m}{\sqrt{a^2m^2+b^2}}, \frac{b^2}{\sqrt{a^2m^2+b^2}}\right) \text{ if } c = \sqrt{a^2m^2+b^2}$$

and $\left(\frac{a^2m}{\sqrt{a^2m^2+b^2}}, \frac{b^2}{\sqrt{a^2m^2+b^2}}\right)$ if $c = \sqrt{a^2m^2+b^2}$

Point 'θ': Any point on the ellipse $\frac{x^2}{a^2}+\frac{y^2}{b^2}=1$ is $(a \cos \theta, b \sin \theta)$ and this point is referred as the point'θ'.

The parametric equation of the ellipse is

$x = a \cos \theta, y = b \sin \theta$

θ is called the ecentric angle.

(c) Hyperbola: Condition that line $y = mx + c$ is tangent to the hyperbola $\frac{x^2}{a^2} - \frac{y^2}{b^2} = 1$ is

$$c = \pm\sqrt{a^2m^2 - b^2}$$

Point of contact is $\left(\frac{ma^2}{\sqrt{a^2m^2 - b^2}}, \frac{b^2}{\sqrt{a^2m^2 - b^2}}\right)$

Point 'θ': ($a \sec\theta$, $b \tan\theta$) **are the coordinates of any point on the hyperbola.**

The parametric equation of hyperbola can be written as $x = a \sec\theta$, $y = b \tan\theta$

16

FREQUENCY TABLES

Statistics: Statistics is the science which deals with the method of collecting and presenting numerical data to throw light on any sphere of enquiry and enables us to interpret it and draw interferences from it.

Characteristics of Statistics:

1. Statistics are the aggregate of facts.
2. Statistics are numerically expressed.
3. Statistics should be placed in relation to each other.
4. Statistics should be collected for a pre-determined purpose.
5. Statistics are affected to a marked extent by multiplicity of causes and not by a single cause.
6. Statistics should be collected in a systematic manner.

7. The reasonable standard of accuracy should be maintained in statistics.

Importance and Usefulness of Statistics:

1. Statistics help in presenting large quantity of data in a simple and classified form.
2. It helps in finding the conditions of relationship between the variables.
3. It gives the methods of comparison of data and it weighs and judges them in the right perspective.
4. It tries to give material for businessman as well as to the administrators so as to serve as a guide in planning and in shaping future policies and programmes.
5. It enlarges individual mind.
6. It when considered as the logic of figures, assists in arriving at, correct views based of facts.
7. It proves useful in a number of fields like Railways, Banks, Army etc.

Raw Data:

The information which we get through censuses, suerveys or in a routine matter is called raw data.

The word 'data' means *information*(or given facts).

The word raw attached to data indicates that this information, thus collected and recorded, cannot be put into use and requires processing.

Collection of Data: The collection of data is the primary need for any statistical investigation. There are two types of data:

Primary Data: It is the data collected by a particular person or organisation for his own use from the primary source.

Mehod of Collecting Primary data:

1. Direct personal observations.
2. Indirect oral investigations.
3. Estimates from local sources and correspondence.
4. Data through questionnaires.
5. Investigation through enumerators.

Secondary Data: It is the data collected by some other person or organisation for their own use but the investigator also gets it for his use.

Method of Collecting the Secondary Data:

1. Information collected through newspaper and periodicals.

2. Information obtained from the research papers published by University department or research bureaus of U.G.C.
3. Information obtained from the official publications of the central, state and the local government dealing with crop statistics, industrial statistics, trade and transport statistics.
4. Informatin obtained from the publications of trade association.
5. Information obtained from the official publications of the foreign government for international organisation.

Classification of Data: There are four bases of classification of data:

1. Quantitative e.g., a class of students split up into groups according to their heights or ages.
2. Qualitative e.g., rich and poor, educated and uneducated persons, intelligent and dull students.
3. Spatial or Geographical (basis of classification is according to difference in geographical location or space) e.g., birth rate of India is divided statewise.
4. Temporal (classification is according to difference in time).

Variables of Observation: Suppose that in a census we record the age, and place of residence (rural or urban) of each person. Here we take observations on three variables: age, sex, and place of residence. Thus, the term variable stands for what is being observed. This is variable because the results are different for different persons. A person may have age 10, another 32, and so on.

Thus 10, 32, ... are the variables. Similarly the values of sex variable are 'male' and 'female'. The age variable can take any value 1, 2, 3, ... but we may not observe all as there may be no person with age 2 at a given time. In view of this we distinguish between 'observed values' and 'possible values'.

Qualitative and Quantitative Variables: In the above example of census, the values of age-variable are numbers, those of sex-variable are names ('male or female') and those variable of place of residence are also name ('rural' or 'urban').

The variables of observation with numbers as possible values are called quantitative variables.

Example: Age, height, income; etc. are quantitative variables. The variables of observation with names of things, place, attributes; etc. as possible values are called quantitative variables.

Example: Sex, religion, caste; etc. quantitative variables.

Units of Observation: The term, unit of observation, is used to describe what the values of a variable are attached to. In the above census example, the units of observation are the persons alive at the time of census and to each unit of observation. We associate, the value of three variables; age, sex and place of residence.

Types of Variable: There are tow types of variable:

(*i*) **Continuous Variable:** A quantity which can take any numerical value within a certain range, is called continuous variable.
Example: Measurement such as height, pressure, temperature; etc. are all continuous variables.

(*ii*) **Discontinuous (Discrete) Variable:** A quantity which is incapable of taking all possible values is called discontinuous variable.

This is also called discrete variable.

Examples:

(I) Number of members in a family can't be $4\frac{1}{2}$.

(II) Marks obtained by a student can't be $47\frac{1}{2}$ etc.

Construction of Frequency Table: If the number of observations in the raw data is large, the investigator has to devise way to condense the data and present them in tables and charts in order to bring out their main features. This is known as data presentation.

Example: The marks obtained by 30 students of XI class in a test (out of 25 marks) in Mathematics are:

17, 5, 18, 17, 18, 2, 16, 13, 8, 17, 8, 18, 2, 13, 17, 8, 16, 18, 8, 5, 13, 8, 18, 16, 8, 13, 18, 2, 5, 18.

These are called ungrouped (or raw) data which do not give any useful information. We can classify them in two ways:

We could, for instance arrange them in ascending or descending order. (This is said to be an array).

But this can not be taken as a rule of the number if observations are large.

We can condense the data into classes (or groups) as below:

(*a*) **Determine the range of raw data** [Range of raw data is the difference between the maximum and the minimum number occurring in the data]

(*b*) **Decide upon the number of classes into which raw data are to be grouped:** [**Rule:** Have no fewer than 5 or more than 12 classes].

The accuracy is lost if we take less number of classes and computation becomes difficult if we take more number of classes. The classes must include the minimum as well as the maximum number occurring in the data.

(*c*) **Size of class-intervals = Range + Desired number of classes.**

In case the quotient is not a positive whole number, then take the next positive whole number as the class size.

(*d*) Set up class limits by using the size of the class interval, keeping in view that the maximum numbers occurring in the data are included in the same class.

Rules:

1. Classes should not be overlapping.
2. There should be no gaps between classes.
3. Classes should be of the same size.

(*e*) Take each number from the data (one at a time) and place a tally mark (/) opposite to the class to which it belongs.

The tally marks can be recorded in bunches of 5.

After the occurrence four times, the fifth ocurrrence is recorded by a cross tally (𝍸) on the first four tallies.

[The counting of tally marks in a particular class is called the frequency of that class].

The table showing how frequencies are distributed is called frequency table.

If the number of observation is less than a particular value, we find cumulative frequency of each class.

(*f*) The table showing the manner in which cumulative frequencies are distributed is called Cumulative Frequency Table.

Relative Frequency Table: Sometimes the frequency is expressed as a fraction of the total frequency and this fraction (usually expressed in percentage) is called the relative frequency.

Consider the frequency inhabitants of U.P. by age (1971 census)

Age	Frequency	%
0–9	26,105,403	29.5
10–14	10,859,933	12.3
15–19	7,184,548	8.1
20–24	6,531,468	7.3
25–29	6,474,870	7.4
30–34	5,910,572	6.7
35–39	5,174,221	5.9
40–44	4,737,880	5.4
45–49	3,676,085	4.2
50–54	3,598,058	4.1
55–59	2,103,947	2.4
60–64	2,636,013	3.0
65–69	1,241,876	1.4
70 and above	2,099,009	2.4

The relative frequencies make it easy to understand when the class frequencies are very large.

Types of Classes: There are two types of classes viz. Exclusive and Inclusive.

(*a*) Consider the classes

5–10, 10–15, 15–20, ...

Here, in the class 5–10, 5 is the lower limit and 10 is the upper limit.

Similar is the case with other classes. Here, the upper limit of a class is the lower limit of the next class in each case.

Thus, there are no gaps.

Size of each class = Its upper limit – Its lower limit

= 10–5 = 15–10 = 20–15= ... = 5.

The common point of two classes is included in the higher class.

Example: $10 \in (10\text{–}15)$ and $10 \notin (5\text{–}10)$

Thus, $x \in (10\text{–}15) \Rightarrow 10 \le x < 15$

[x can't = 15]

Thus, the upper limit is excluded.

Such classes are called exclusive classes.

(*b*) Consider the classes

10–19, 20–29, 30–39, ...

$\therefore$ $20 \in (20\text{–}29)$ and $29 \in (20\text{–}29)$.

Here $x \in (20–29) \Rightarrow 20 \le x \le 29$

Thus, the upper limit is included.

Such classes are called inclusive classes.

Sum up:

The lower limit is always inclusive.

The class is inclusive or exclusive according as the upper limit is included or excluded.

Method to convert inclusive classes into exclusive classes?

We call

$$\left.\begin{array}{l}\textbf{adjustment}\\ \textbf{factor}\end{array}\right\} = \frac{\begin{array}{c}\text{Lower limit}\\ \text{of 2nd class}\end{array} - \begin{array}{c}\text{Upper limit}\\ \text{of 1st class}\end{array}}{2}$$

Subtract this adjustment factor from the lower limits and add to the upper limits so as to get the required exclusive classes.

Example: The inclusive classes are 10–19, 20–29, 30–39, 40–49, ...

$$\textbf{Adjustment factor} = \frac{20-19}{2} = \frac{1}{2} = .5$$

Subtracting .5 from the lower limits and adding to upper limits, we get

9.5–19.5, 19.5–29.5, 29.5–39.5, 39.5–49.5,....

These are exclusive classes.

The mid-point of a class-interval is called its Class-mark or Mid-value.

Example: class-mark of 9.5–19.5 is

$$\frac{9.5+19.5}{2}=\frac{29}{2}=14.5$$

Remember:

$$\text{Class-mark}=\frac{\text{Lower limit}+\text{Upper limit}}{2}$$

Graphical Presentation of Frequency Distribution: The graphical representation of the frequency distribution is convenient communication method.

(*i*) **Simple Bar Diagram:** It is used to compare two or more items related to variable. In this case the data are presented with the help of bars. These bars are usually arrranged according to relative magnitude of bars. The length of a bar is determined by the value or the amount of the variable. A limitation of Simple Bar Diagram is that only one variable can be represented on it.

Illustration: Draw a Bar chart of the procurement of rice (in tons) in an Indian state:

Year:	1978	1979	1980	1981	1982	1983
Rice (in tons):	4500	5700	6100	6500	4300	7800

Solution: The given data is represented by the Simple Bar Diagram as only one variable is to be presented.

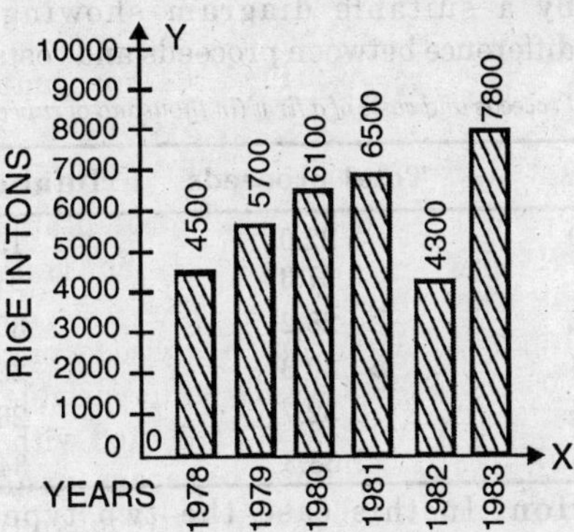

Here we represent years on the x-axis and procurement of rice on the y-axis.

(*ii*) **Multiple or Grouped Bar Diagram:** A multiple or grouped bar diagram is used when a number of items are to be compared in respect of two, three or more values. In this case the numerical values of major categories are arranged in ascending or descending order so that the categories can be readily distinguished. Different shades or colours are used for each category.

Illustration: Represent the following data by a suitable diagram showing the difference between proceeds and costs:

Proceeds and costs of a firm (in thousand of rupees)

Year	Total proceeds	Total costs
1950	22.0	19.5
1951	27.3	21.7
1952	28.2	30.0
1953	30.3	25.6
1954	32.7	26.1
1955	33.3	34.2

Solution: In this case the two types of information, proceeds and costs, are shown on a diagram indicating the difference between them. The two types of information for each year are placed in such a way that a comparison may be made between them.

So, a grouped bar diagram is drawn in order to represent the given data.

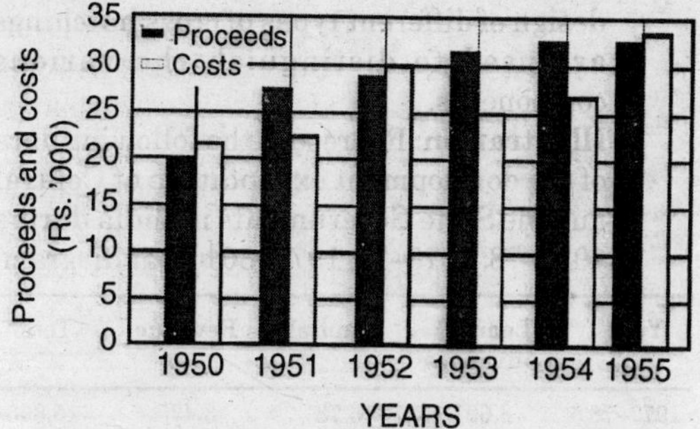

(*iii*) **Subdivided or Component Bar Diagram:** A component bar diagram is one which is formed by dividing a single bar into several component parts. A simple bar represents the aggregate value whereas the component parts represent the component values of the aggregate value. It shows the relationship among the different parts and also between the different parts and the main bar. The design and procedure of constructing it is similar to simple bar diagrams except that in this form of presentation each bar is subdivided into its components. Different shades or colours or design of different types of cross hatchings are used to distinguish the various components.

Illustration: Represent the following data of the development expenditure of Central and the State Governments in India during 1977–78, 1978–79, 1979–80 by bar diagram.

Year	Loans & Advances	Capital	Revenue	Total
1977–78	8,601	3,787	3,477	15,865
1978–79	10,535	4,456	4,036	18,827
1979–80	11,549	4,803	3,709	20,061

Solution: In this case we use subdivided bar diagram. Here the data for different years is represented by various parts of the single bar for the years. Different parts are indicating loans and advances; Capital and revenues for each year as shown in the figure.

(*iv*) **Percentage Sub-divided Bar Diagrams:** It consists of one or more than one bars where each bar totals 100%. Its construction is similar to the subdivided bar diagram with the only difference that whereas in the subdivided bar diagram segments are used in absolute quantities, in the percentage

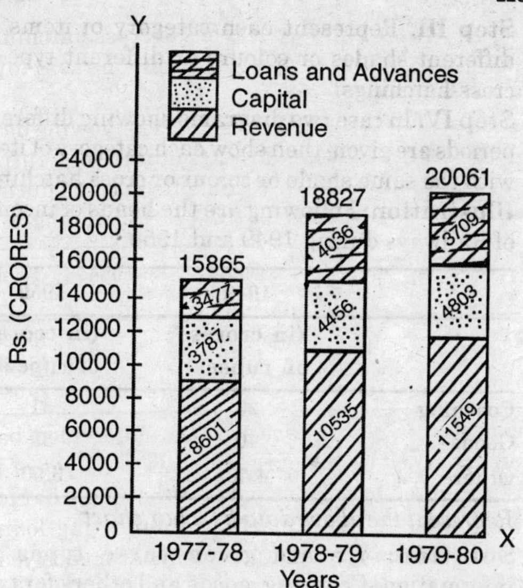

bar diagram the quantities are transformed into percentages. Its construction is based on the following steps:

Step I. Convert the quantities in each case into percentage of the whole

Step II. Take the cumulative percentages.

Step III. Represent each category of items in different shades or colours or different type of cross-hatchings.

Step IV. In case two diagrams showing different periods are given, then show each category of item with the same shade or colour or cross hatching.

Illustration: Following are the heads of income of Railways during 1949 and 1950.

	1949	1950
	(in crores of rupees)	(in crores of rupees)
Coaching	26	31
Goods	40	39
Others	4.50	3.50

Represent the above data by a bar chart.

Solution: We are given three types of information–Coaching, goods and others for two years. In order to facilitate comparison among them and also between the two years, a component bar chart is drawn to represent the given data. The graph has been drawn on percentage figures. Percentage of each item has been expressed on the total income of respective years by formula

$$\text{Percentage} = \frac{\text{Income of the item}}{\text{Total income of respective year}} \times 100$$

Calculation of percentage.

Items	1949		1950	
	Income	Percentage	Income	Percentage
Coaching	26	36.88	31	42.18
Goods	40	56.74	39	53.06
Other	4.50	6.38	3.50	4.76
	70.50	100	73.50	100

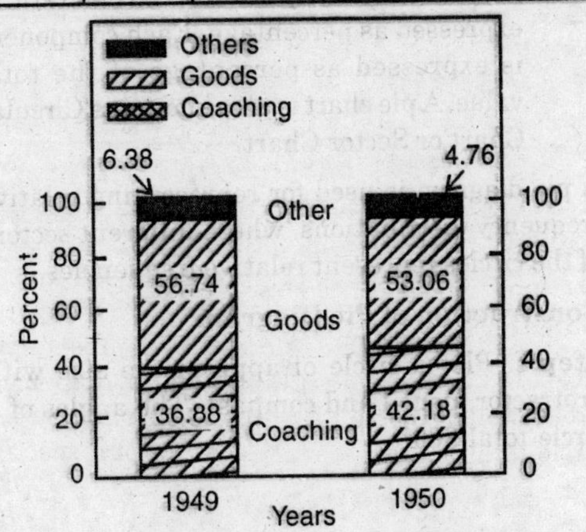

(*v*) **Pie Chart or Sector Chart:** A pie chart is a circular graph which represents the total value with its components. The area of a circle represents the total value and the different sectors of the circle represent the different part. The circle is divided into sectors by radii and the areas of the sectors are proportional to the angles at the centre. It is generally used for comparing the relation between various components of a value and between components and the total value. In pie chart, the data is expressed as percentage. Each component is expressed as percentage of the total value. A pie chart is also known as Circular Chart or Sector Chart.

A pie diagram is used for representing relative frequency distributions, where different sectors of the circles represent relative frequencies.

Construction of Pie Diagram

Step I. Plot a circle of appropriate size with protractor, pencil and compass. The angles of a circle total 360°.

Step II. Convert the given value of the components of an item in percentage of the total value of the item.

Step III. In laying out the sector for a pie chart it is logical to adopt the common procedure to arrange sectors according to size with the largest at the top and the others in sequence running clockwise.

Step IV. Transpose the various component values corresponding to the degree on the circle. Since 100% is represented by 360° angle at the centre of the circle, therefore 1% value is represented by

$$\frac{360^\circ}{100} = 3.6^\circ.$$

If 8 be the percentage of a certain component, the angle which represents the percentage of such component is (3.6×8) degrees.

Setp V. Measure with protractor the points on a circle representing the size of each sector.

Step VI Label each sector for identification.

Illustration: Draw a pie chart to represent the following data on the proposed out lay during the Fourth Five Year Plan.

Items	Agriculture	Industries and Minerals	Irrigation and Power	Communication	Miscellaneous
Rs. in Crores	6000	4000	2500	4500	3000

Solution: In constructing a pie chart, it is necessary to convert the percentages into angles of different degrees.

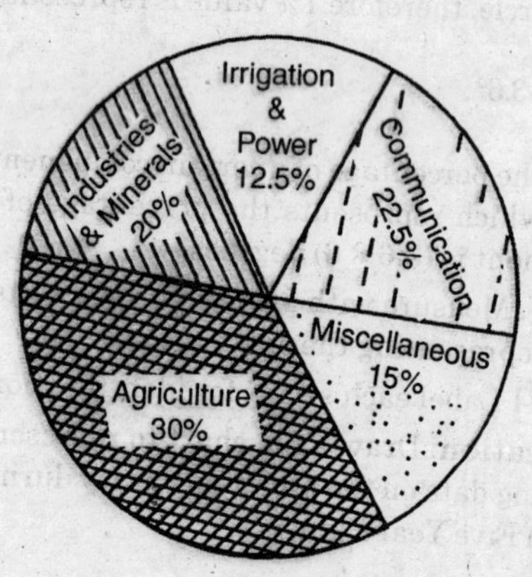

Calculation for Pie Chart

Items	Amount in (Rs.) Crores	Percentage on total (%)	Angle for each percentage (360° ÷ 100)	Angle for each item at the centre of the Pie-chart
Agriculture	6000	$\frac{6000 \times 100}{20000} = 30$	3.6°	30 ×3.6° = 108°
Industries and Minerals	4000	$\frac{4000 \times 100}{20000} = 20$	3.6°	20 ×3.6° = 72°
Irrigation and Power	2500	$\frac{2500 \times 100}{20000} = 12.5$	3.6°	12.5 × 3.6° = 45°
Communication	4500	$\frac{4500 \times 100}{20000} = 22.5$	3.6°	22.5 × 3.6° = 81°
Miscellaneous	3000	$\frac{3000 \times 100}{20000} = 15$	3.6°	15 × 3.6° = 54°
Total	20000	100		360°

Arithmetic Mean

(*i*) **Mean of Raw Data:** We know that the average of n-numbers is obtained by finding their sum (by adding) and then dividing it by n. Let $x_1, x_2, x_3, \ldots, x_n$ be n-numbers, then their average is given by

$$\bar{x} = \frac{x_1 + x_2 + x_3 \ldots + x_n}{n}$$

Illustration: Find the arithmetic mean of the marks obtained by 10 students of class X in Mathematics in a certain examination. The marks obtained are: 25, 30, 21, 55, 47, 10, 15, 17, 45, 35.

Solution: Let $\bar{x}$ be the average marks

$\therefore$ Sum of all the observations

$= 25 + 30 + 21 + 55 + 47 + 10 + 15 + 17 + 45 + 35 = 300$

Number of students = 10

$\therefore$ Arithmetic mean $= \frac{300}{10} = 30.$

(*ii*) **Mean of Grouped Data:**

Let $x_1, x_2, x_3, \ldots x_n$ be the variates and let $f_1, f_2, f_3, \ldots f_n$ be their corresponding frequencies, then their mean $\bar{x}$ is given by

$$\bar{x} = \frac{f_1x_1 + f_2x_2 + f_3x_3 + \ldots\ldots f_nx_n}{f_1 + f_2 + f_3 + \ldots\ldots + f_n} = \frac{\sum_{i=1}^{n} f_ix_i}{\sum f_i} = \frac{\sum_{i=1}^{n} f_ix_i}{N}$$

where $N = f_1 + f_2 + f_3 + \ldots. + f_n$

Illustration: Find the Arithmetic Mean from the frequency table

Marks	52	58	60	65	68	70	75
No. of Students	7	5	4	6	3	3	2

Solution: Let x be the marks and f be the frequency so that we have the following table:

x	f	fx
52	7	364
58	5	290
60	4	240
65	6	390
68	3	204
70	3	210
75	2	150
Total	30	1848

Here $N = \Sigma f = 30$ and $\Sigma fx = 1848$

$$\therefore \text{ Mean} = \bar{x} = \frac{1848}{30} = \frac{616}{10} = 61.6$$

Short-cut Method

(*i*) In the case of ungrouped data

$$\bar{x} = a + \frac{\Sigma d}{n}, \text{ where}$$

a = assumed mean, n = number of items,
$d = x - a$ = deviations of any variate from a.

Working Rule for Short Cut Method for Ungrouped Data

Step I. Denote the variable of the discrete series by x or X.

Step II. Take any item of series, preferable the middle one, and denote it by a. This number a is called the assumed mean or provision mean.

Step III. Take the difference $x - a$ and denote it by d or dx or $d' = x - \text{a}$, where d' is the deviation of variate from 'a'.

Step IV. Find the sum Σdx or Σd.

Step V. Use the following formula to calculate the arithmetic mean.

$$\bar{x} = a + \frac{\Sigma d}{n}$$

Illustration: Find, by short-cut method, the mean height of the following 8 students whose heights in centimetre are

59, 65, 71, 67, 61, 63, 69, 73.

Solution: Let us take 65 as assumed mean i.e., $a = 65$. Let us prepare the following table.

x	$d = x - 65$
59	–6
61	–4
63	–2
65	0
67	+2
69	+4
71	+6
73	+8
	$\Sigma d = 8$

Total deviation = $\Sigma d = 8$

Here $a = 65$, $n = 8$, $\Sigma d = 8$.

$$\therefore \quad \bar{x} = a + \frac{\Sigma d}{n} = 65 + \frac{8}{8} = 66 \text{ cm.}$$

(*ii*) In the case of grouped data $\bar{x} = a + \frac{\Sigma fd}{n}$

where fd = product of the frequency and the corresponding deviation.

$N = \Sigma f$ = the sum of all the frequencies.

Working Rule for short-cut Method for Grouped Data

Step I. In the case of discrete series, denote the variable by x or X and the corresponding frequency by f. (But in the case of continuous series x is the mid-value of the interval and f, the frequency corresponding to that interval.)

Step II. Take any item of x series, preferable the middle one and denote it by 'a'. This number 'a' is called assumed mean or provisional mean.

Step III. Take the difference $x - a$ and denote it by d or dx or $d' = x - a$ = deviation of any variate x from a, the assumed mean.

Step IV. Multiply the respective f and d and denote the product under the column fd.

Step V. Find Σfd.

Step VI. Use the following formula to calculate the arithmetic mean.

$$\bar{x} = a + \frac{\Sigma fd}{\Sigma f}$$

Illustration. Ten coins were tossed together and the number of tails resulting from them were observed. Calculate the A.M.

x:	0	1	2	3	4	5	6	7	8	9	10
f:	2	8	43	133	207	260	213	120	54	9	1

Solution: Let 5 be the assumed mean i.e., $a = 5$.

x	f	$d = x - 5$	fd
0	2	– 5	– 10
1	8	– 4	– 32
2	43	– 3	– 129
3	133	– 2	– 266
4	207	– 1	– 207
5	260	0	0
6	213	1	+ 213
7	120	2	+ 240
8	54	3	+ 162
9	9	4	+ 36
10	1	5	+ 5
	$\Sigma f = 1050$		$\Sigma fd = 12$

$$\therefore \text{A.M.} = \bar{x} = a + \frac{\Sigma fd}{\Sigma f} = 5 + \frac{12}{1050} = 5 + 0.0114$$

$$= 5.0114$$

Illustration: For the following frequency table, find the mean class:

100–120	120–140	140–160	160–180	180–200	200–220	220–240

Frequency:

10	8	4	4	3	1	2

Solution: Let $a = 170$, then we have the following table.

Class	Frequency f	Mid-value x	$d = x - a$ $= x - 170$	fd
100–120	10	110	– 60	– 600
120–140	8	130	– 40	– 320
140–160	4	150	– 20	– 80
160–180	4	170	0	0
180–200	3	190	20	60
200–220	1	210	40	40
220–240	2	230	60	120
	$\Sigma f = 32$			$\Sigma fd = -780$

$$\text{Mean } = \bar{x} = a + \frac{\Sigma fd}{\Sigma f} = 170 - \frac{780}{32}$$

$$= 170 - 24.3 = 145.7$$

Step Deviation Method

When the class intervals in a grouped data are equal, then the calculations can be simplified further by taking out the common factor from the deviations.

$$A.M. = \bar{x} = a + \frac{\Sigma fd}{N} \times i$$

where a = assumed mean; $d = \frac{x-a}{i}$ = the deviation of any variate from a; i = the width of the class-interval, N = Number of observations.

Illustration: Find the Arithmetic mean of the followig data:

10–20	20–30	30–40	40–50	50–60	60–70	70–80	80–90	90–100
Frequency:								
2	7	17	29	29	10	3	2	1

Solution: Let the assumed mean $a = 45$

Class Interval	Mid.value x	$d=\frac{x-45}{10}$	f	fd
10–20	15	– 3	2	– 6
20–30	25	– 2	7	– 14
30–40	35	– 1	17	– 17
40–50	45	0	29	0
50–60	55	1	29	29
60–70	65	2	10	20
70–80	75	3	3	9
80–90	85	4	2	8
90–100	95	5	1	5
			N=100	Σfd=34

$a = 45$, N = 100, $\Sigma fd = 34$ and $i = 10$

$$\bar{x} = a + \frac{\Sigma fd}{\Sigma f} \times i$$

$$= 45 + \frac{34}{100} \times 10 = 45 + 3.4 = 48.4$$

Weighted Arithmetic Mean

If $w_1, w_2, w_3, \ldots, w_n$ are the weights assigned to the values $x_1, x_2, x_3, \ldots, x_n$ respectively, then the weighted average is defined as:

Weighted Arithmetic Mean

$$= \frac{w_1x_1 + w_2x_2 + w_3x_3 + \ldots + w_nx_n}{w_1 + w_2 + w_3 + \ldots + w_n}$$

Median: Median is defined as the middle most or the central value of the variables in a set of observations, when the observations are arranged either in ascending or descending order of their magnitudes. It is denoted by M.

Calculation of Median

(*a*) **When the data is ungrouped:** Arrange the n values of the variable in ascending (or descending) order of magnitudes.

Case I. When n is odd: In this case $\frac{n+1}{2}$th value is the median

i.e., $M = \frac{n+1}{2}$th term.

Illustration: The number of runs scored by 11 players of a cricked team of a school are 5, 19, 42, 11, 50, 30, 21, 0, 52, 36, 27. Find the median.

Solution: Let us arrange the values in ascending order

0, 5, 11, 19, 21, 27, 30, 36, 42, 50, 52 ...(1)

$$\therefore \text{Median, } M = \left(\frac{n+1}{2}\right) \text{th value}$$

$$= \left(\frac{11+1}{2}\right) \text{th value} = 6\text{th value.}$$

Now the 6th value in the data (1) is 27,

$\therefore$ Median = 27 runs.

Case II. When *n* is even: In this case there are two middle terms $\frac{n}{2}$ th and $\left(\frac{n}{2}+1\right)$ th. The median is the average of these two terms *i.e*,.

$$M = \frac{\frac{n}{2} + \left(\frac{n}{2}+1\right)}{2}$$

Illustration: Find the median of the following item: 6, 10, 4, 3, 9, 11, 22, 18.

Solution: Let us arrange the items in ascending order, 3, 4, 6, 9, 10, 11, 18, 22
In this data the number of items is $n = 8$, which is even.

$$\therefore \text{Median} = M = \text{average of } \left(\frac{n}{2}\right)\text{th and } \left(\frac{n}{2}+1\right)\text{th terms}$$

$$= \text{Average of } \left(\frac{8}{2}\right)\text{th and } \left(\frac{8}{2}+1\right)\text{th terms}$$

= Average of 4th and 5th terms

$$= \frac{9+10}{2} = \frac{19}{2} = 9.5$$

(*b*) **When the data is grouped:**

(i) **When the series is discrete:** In this case the values of the variable are arranged in ascending or descending order of magnitudes. A table is prepared showing the corresponding frequencies and cumulative frequencies. Then the median is calculated as:

$$M = \left(\frac{n+1}{2}\right)\text{th value}$$

where n = Σf = total frequencies.

Illustration: Calculate median for the following data:

No. of Students:	6	4	16	7	8	2
Marks :	20	9	25	50	40	80

Solution: Arranging the marks in ascending order and preparing the following table.

Marks	Frequency	Cumulative Frequency
9	4	4
20	6	10
25	16	26
40	8	34
50	7	41
80	2	43
	$n = \Sigma f = 43$	

Here, $n = 43$

$$\therefore \text{Median } = M = \left(\frac{n+1}{2}\right)\text{th value.}$$

$$= \left(\frac{43+1}{2}\right)\text{th value} = 22\text{nd value}$$

The above table shows that all items from 11 to 26 have their values 25. Since 22nd item lies in this interval, therfore its value is 25.

Hence, median = 25.

(c) **When the series is continuous.** In this case the data is given in the form of a frequency table with class-interval etc., and the following formula is used to calculate the median.

$$M = L + \frac{\frac{n}{2} - C}{f} \times i, \text{ where}$$

L = lower limit of the class in which the median lies

n = total number of frequencies, i.e., $n = \Sigma f$

f = frequency of the class in which the median lies

C = cumulative frequency of the class proceeding the median class

i = width of the class-interval of the class in which the median lies.

Illustration: The following table gives the marks obtained by students in Economics. Find the median.

Marks	No. of Students
10–14	4
15–19	6
20–24	10
25–29	5
30–34	7
35–39	3
40–44	9
45–49	6

Solution: Let us prepare the following table showing the frequencies and cumulative frequencies.

Marks	Frequency	Cumulative Frequencies
10–14	4	4
15–19	6	10
20–24	10	20
25–29	5	25
30–34	7	32
35–39	3	35
40–44	9	44
45–49	6	50

Here $n = 50$, $\therefore \frac{n}{2} = 25.$

It is evident that 25th item is the median and in the table the cumulative frequency which contains the 25th item lies in the interval 25 –29.

Median class = 25 – 29

Now the given series is inclusive one. Let us convert it to exclusive series and hence the value of L will be $\frac{25+24}{2} = 24.5$

Also, $\frac{n}{2} = 25$

$\therefore$ L = lower limit of the median class = 24.5

C = cumulative frequency of the class (20–24) preceding the median class = 20

f = frequency of the median class = 5

i = class-interval of the median class = 5

$$\therefore \text{Median} = 24.5 + \frac{25-20}{5} \times 5$$

$$= 24.5 + 5 = 29.5$$

Measures of Dispersion. Dispersion means scatterness. The degree to which numerical data tend to spread about an average value is called the dispersion of the data. The term dispersion is used in two senses. The first relates to the limits within which the data fall and the second takes into account the amount, absolute or relative, by which the value of item differ from an average.

Mean Deviation: It is the average of the modulus of the deviations of the observations in a series taken from mean or median.

Methods for Calculation of Mean Deviations

(*a*) **For ungrouped data:** In this case the mean deviation is given by the formula

$$\text{Mean deviation} = \frac{\Sigma|x - A|}{n} = \frac{\Sigma|d|}{n}$$

where '*d*' stands for the deviation from the mean or median and | d | is always positive whether *d* itself is positive or negative and *n* is the total number of items.

Illustration: Find the mean deviation from (*i*) mean and (*ii*) median for the following data:

Marks	: 20	18	16	14	12	10	8	6
No. of Students:	2	4	9	18	27	25	14	1

Solution: (*i*) **Mean deviation from mean:** Let us calculate the mean of the given data by forming the following table:

Marks (x)	No. of Students (f)	$f \times x$
6	1	6
8	14	112
10	25	250
12	27	324
14	18	252
16	9	144
18	4	72
20	2	40
	$\Sigma f = 100$	$\Sigma f \times x = 1200$

$\therefore$ Arithmetic mean $= \dfrac{1200}{100} = 12$

We shall now proceed to calculate the mean deviation. Let us prepare the following table for it.

(x)	(f)	$\lvert d \rvert = \lvert x - 12 \rvert$	$f \lvert d \rvert$
6	1	6	6
8	14	4	56
10	25	2	50
12	27	0	0
14	18	2	36
16	9	4	36
18	4	6	24
20	2	8	16
	$\Sigma f = 100$		$\Sigma f \lvert d \rvert = 224$

$\therefore$ Mean deviation (about mean)

$$= \frac{\Sigma f \lvert d \rvert}{n} = \frac{224}{100} = 2.24$$

(*ii*) **Mean deviation about median:** Let us prepare the following table in order to calculate the median:

Marks (x)	Frequency	Cumulative frequency
6	1	1
8	14	15
10	25	40
12	27	67

14	18	85
16	9	94
18	4	98
20	2	100

Now Median = Average of $\frac{n}{2}$th and $\left(\frac{n}{2}+1\right)$th item.

= Average of 50th and 51st item = 12

Let us now prepare the following table in order to calculate the mean deviation from median:

x	f	$\lvert d\rvert = \lvert x-12\rvert$	$f\times\lvert d\rvert$
6	1	6	6
8	14	4	56
10	25	2	50
12	27	0	0
14	18	2	36
16	9	4	36
18	4	6	24
20	2	8	16
	$\Sigma f = 100$		$\Sigma f\times\lvert d\rvert = 224$

$$\therefore \text{Mean deviation (about median)} = \frac{224}{100} = 2.24$$

(*b*) **Mean Deviation for grouped data:** Let $x_1, x_2, x_3, ... x_n$ occur with frequency $f_1, f_2, f_3, ... f_n$ respectively and let $\Sigma f = n$ and M can be either Mean or Median then the mean deviation is given by the formula.

$$\textbf{Mean Deviation} = \frac{\Sigma f|x-M|}{\Sigma f} = \frac{\Sigma f|d|}{n}$$

where $d = | x - M |$ and $\Sigma f = n$.

Illustration: Calculate the mean deviation from the mean for the following data:

Class-interval :	0–4	4–8	8–12	12–16	16–20
Frequency :	4	6	8	5	2

Solution: Let us prepare the following table by assuming that the frequencies in each class are centred at its mid-value.

Class-interval	Mid-value	f	fx	$\|d\|$ $= \|x-9.2\|$	$f\|d\|$
0–4	2	4	8	7.2	28.8
4–8	6	6	36	3.2	19.2
8–12	10	8	80	0.8	6.4
12–16	14	5	70	4.8	24.0
16–20	18	2	36	8.8	17.6
	$\Sigma f = 25$		$\Sigma fx = 230$	$\Sigma f\|d\| = 96.0$	

Now A.M. $= \frac{\Sigma fx}{\Sigma f}$, $\therefore$ A.M. $= \frac{230}{25} = 9.2$

Also, Mean Deviation (about mean) $= \frac{\Sigma f|d|}{\Sigma f}$

$\therefore$ Mean deviation $= \frac{96}{25} = 3.84$

Merits and Demerits of Mean Deviation

Merits:
1. It is easy to understand and compute.
2. Mean deviation is less affected by the extreme values as compared to standard deviation.
3. Mean deviation about an arbitrary point is least when the point is median.

Demerits:
1. In mean deviation the signs of all deviations are taken as positive and therefore, it is not suitable for further algebraic treatment.
2. It is rarely used in social sciences.
3. It does not give accurate results because the mean deviation from the median is least but median itself is not considered a satisfactory average when the variation in the series is large.

Standard Deviation: The positive square root of the average of squared deviations of all observations taken from their mean is called standard deviation. It is generally denoted by the Greek alphabet σ or s.

Variance: The square of the standard deviation is called variance and is denoted by σ^2.

Coefficient of Standard Deviation: It is the ratio of the standard deviation to its A.M. *i.e.*,

$$\text{Coefficient of standard deviation} = \frac{\sigma}{\bar{x}}$$

Coefficient of Variance: It is the product of coefficient of standard deviation × 100.

$$\text{Coefficient of variance} = \frac{\sigma}{\bar{x}} \times 100.$$

Computation of Standard Deviation: The methods of calculating the standard deviation depend upon the nature of data and also on the number of observations for ungrouped data.

(*a*) **Standard Deviation for Ungrouped Data:**

Direct method: In case of simple series, the standard deviation can be obtained by the formula:

$$\sigma = \sqrt{\frac{(x-\bar{x})^2}{n}} = \sqrt{\frac{\Sigma d^2}{n}}, \text{ where } d = x - \bar{x}$$

and x = value of the variable or observation, $\bar{x}$ = arithmetic mean, n = total number of observations.

Illustration: Find the standard deviation of 16, 13, 17, 22.

Solution: Here

$$\text{A.M.} = \bar{x} = \frac{16+13+17+22}{4} = \frac{68}{4} = 17.$$

Let us prepare the following table in order to calculate the standard deviation.

(x)	$d = x - \bar{x} = x - 17$	$(x-\bar{x})^2$
16	– 1	1
13	– 4	16
17	0	0
22	5	25
		$\Sigma d^2 = 42$

Now $\sigma = \sqrt{\frac{\Sigma(x-\overline{x})^2}{n}} = \sqrt{\frac{42}{4}} = 3.2$

Short-cut Method: This method is applied to calculate standard deviation, when the mean of the data comes out to be a fraction. In that case it is very difficult and tedious to find the deviations of all observations from the mean by the earlier method. The formula used is

$$\sigma = \sqrt{\frac{\Sigma d^2}{n} - \left(\frac{\Sigma d}{n}\right)^2},$$

where $d = x - A$, A = assumed mean, n = total number of observations.

Illustration: Find the standard deviation of the following data:

48, 43, 65, 57, 31, 60, 37, 48, 59, 78.

Solution: Let us prepare the following table in order to calculate the value of S.D:

Value (x)	$d = x - A, (A = 50)$	d^2
48	– 2	4
43	– 7	49
65	15	225
57	7	49
31	– 19	361
60	10	100
37	– 13	169
48	– 2	4
59	9	81
78	28	784
$n = 10$	$\Sigma d = 26$	$\Sigma d^2 = 1826$

Here, $\bar{x} = A + \frac{\Sigma d}{n} = 50 + \frac{26}{10} = 52.6$, which is fraction. Let us apply the short-cut formula in order to calculate S.D.

$$\therefore \text{ S.D.} = \sigma = \sqrt{\frac{\Sigma d^2}{n} - \left(\frac{\Sigma d}{n}\right)^2} = \sqrt{\frac{1826}{10} - \left(\frac{26}{10}\right)^2}$$

$$= \sqrt{182.60 - 6.76} = \sqrt{175.84} = 13.26$$

(b) Standard Deviation for Grouped Data or Discrete Series.

Direct Method: The standard deviation for the discrete series is given by the formula.

$$\sigma = \sqrt{\frac{\Sigma f(x-\bar{x})^2}{n}},$$

where $\bar{x}$ is A.M., x is the size of the item, and f is the corresponding frequency in the case of discrete series. But when the mean has a fractional value, then the following formula is applied to calculate S.D.

$$\sigma = \sqrt{\frac{\Sigma fd^2}{n} - \left(\frac{\Sigma fd}{n}\right)^2}$$

where $d = x - A$, A = assumed mean, $n = \Sigma f$ = Total frequency.

Illustration: Find the standard deviation from the following data:

Size of the item	:	10	11	12	13	14	15	16
Frequency	:	2	7	11	15	10	4	1

Also, find the coefficient of variation.

Solution: Let us prepare the following table:

Size of the (x) item	*Frequency f*	*d = x – A, A = 13*	*fd*	*fd^2*
10	2	– 3	– 6	18
11	7	– 2	– 14	28
12	11	– 1	– 11	11
13	15	0	0	0
14	10	1	10	10
15	4	2	8	16
16	1	3	3	9
Total	$n = \Sigma f = 50$		$\Sigma fd = -10$	$\Sigma fd^2 = 92$

$$\text{Now A.M.} = \bar{x} = A + \frac{\Sigma fd}{n} = 13 + \frac{(-10)}{50} = 12.8$$

Here $\bar{x} = 12.8$ is a fraction,

$$\therefore \text{ S.D.} = \sigma = \sqrt{\frac{\Sigma fd^2}{n} - \left(\frac{\Sigma fd}{n}\right)^2} = \sqrt{\frac{92}{50} - \left(-\frac{10}{50}\right)^2}$$

$$= \sqrt{1.84 - 0.04} = \sqrt{1.80} = 1.342$$

$\therefore$ Now the coefficient of variation

$$= \frac{\sigma}{x} \times 100 = \frac{1.342}{12.8} \times 100 = 10.4$$

(c) Standard Deviation in Continuous Series:

Direct Method: The standard deviation in the case of continuous series is obtained by the following formula

$$\sigma = \sqrt{\frac{\Sigma f(x-\bar{x})^2}{n}},$$

where x = mid-value, $\bar{x}$ = A.M., f = frequency, n = total frequency.

Illustration: Calculate the standard deviation for the following frequency distribution.

Class Interval :	0–4	4–8	8–12	12–16
Frequency :	4	8	2	1

Solution: Assuming that frequency at each class is centred at its mid-value, let us prepare the following table:

Table for Arithmetic Mean

Class Interval	Mid-value (x)	Frequency f	Product $f \times x$
0–4	2	4	8
4–8	6	8	48
8–12	10	2	20
12–16	14	1	14
		$\Sigma f = 15$	$\Sigma fx = 90$

$$\text{A.M.} = \bar{x} = \frac{\Sigma fx}{\Sigma f} = \frac{90}{15} = 6$$

Table for Standard Deviation

Mid-value x	Frequency f	$x - \bar{x}$ $= x - 6$	$(x - \bar{x})^2$ $= (x - 6)^2$	$f(x - \bar{x})^2$ $= f(x - 6)^2$
2	4	– 4	16	64
6	8	0	0	0
10	2	4	16	32
14	1	8	64	64
	$\Sigma f = 15$			$\Sigma f(x - \bar{x})^2 = 160$

$$\therefore \text{SD} = \sqrt{\frac{\Sigma f(x-x)^2}{n}} = \sqrt{\frac{160}{15}} = \sqrt{10.67} = 3.27$$

Short Method: The formula for short method to find the standard deviation of continuous series is

$$\sigma = \sqrt{\frac{\Sigma fd^2}{n} - \left(\frac{\Sigma fd}{n}\right)^2} \times i \text{ ,where } d = \frac{x - A}{i},$$

A = Assumed mean, n = total frequency,
i = class width.

Illustration. Find the standard deviation for the following distribution.

Marks	10–20	20–30	30–40	40–50	50–60	60–70	70–80
No. of Students	5	12	15	20	10	4	2

Solution: Let us prepare the following table in order to calculate the standard deviation.

Mark (Class interval)	No. of students (f)	Mid-value (x)	$d = \frac{x-45}{10}$	fd	fd^2
10–20	5	15	–3	–15	45
20–30	12	25	–2	–24	48
30–40	15	35	–1	–15	15
40–50	20	45	0	0	0
50–60	10	55	1	10	10
60–70	4	65	2	8	16
70–80	2	75	3	6	18
Total	$\Sigma f = n$ = 68			Σfd = –30	Σfd^2 = 152

$$\therefore \sigma = i \times \sqrt{\frac{\Sigma fd^2}{n} - \left(\frac{\Sigma fd}{n}\right)^2} = 10 \times \sqrt{\frac{152}{68} - \left(\frac{-30}{68}\right)^2}$$

= 14.3 Approx.

Merits and Demerits of Standard Deviation

Merits:

1. It is based on all the observations.
2. It is rigidly defined.
3. It lend itself to further algebraic treatment.
4. It is less affected by fluctuations of sampling as compared to other measures of dispersion.
5. It is extremely useful in correlation.
6. Like mean deviation, there is no artificiality in it.

Demerits:

1. It is difficult to compute unlike other measures of dispersion.
2. It is not simple to understand.
3. It gives more weightage to extreme values.

LINEAR PROGRAMMING

Linear Inequations in Two Variables: A statement of any one of the following types :

(*i*) $ax + by + c > 0$

(*ii*) $ax + by + c \geq 0$

(*iii*) $ax + by + c < 0$

(*iv*) $ax + by + c \leq 0$

where $a, b, c \in R$ and $a^2 + b^2 \neq 0$, is called a linear inequation (for inequality) in two variables x and y.

An ordered pair (α, β) of real numbers may or may not satisfy a given inequation. The set of all ordered pairs, which satisfy a given inequation, is called the solution set of the given inequation. Also, we know that there is one-one correspondence between the ordered pairs of real numbers and the points of the co-ordinated plane, therefore, it is possible to represent the solution set of a given inequation (in two variables) by the points of a co-ordinated plane. The set of all points whose co-ordinates satisfy a given inequation is called the graph of the inequation.

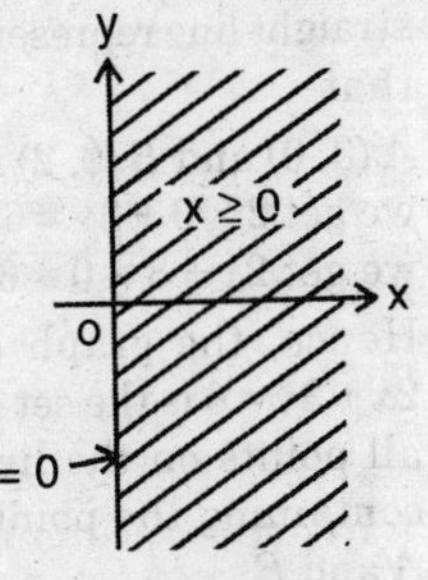

Let us consider the inequaliy $x \geq 0$. The set of this inequality is $\{(x, 0) : x \in R, x \geq 0\}$ and hence the graph of $x \geq 0$ is the set of all points of the *XOY*-plane whose abscissae (*i.e.*, *x*-axis) are non-negative *i.e.*, the set of all points which lie one *Y*-axis or on the right hand side of *Y*-axis. The graph of $x \geq 0$ is shown shaded in fig.

In general, to find the graph of an inequation, we note that $ax + by + c = 0$ represents a straight line, which divides the *XOY*-plane into two halves, one half is the graph of $ax + by + c > 0$ and the other of $ax + by + c < 0$. The illustrate the method of finding the graph of an inequation, we consider the following illustration.

Illustration: Draw the graph of the inequation $2x + 3y \geq 6$

Solution: The give inequality is

$$2x + 3y \geq 6 \qquad ...(1)$$

Consider the equation $2x + 3y = 6$. To find the straight line represented by $2x+3y=6$, we observe that

A (3, 0) and B (0, 2) lie on the line (Putting $x = 0$, we get $2 \times 0 + 3y = 6 \Rightarrow y = 2$ and on putting $y = 0$, we get $2x + 3 \times 0 = 6 \Rightarrow x = 3$).

Hence, the graph of $2x+3y=6$ is the set of all points on the line containing the points A and B.

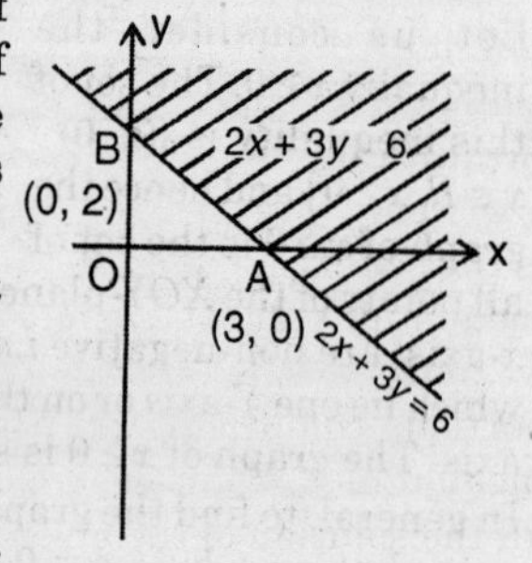

Further we notice that O (0, 0), which lies below the line AB does not satisfy $2x+3y \geq 6$, [$\because 2\times0+3\times0=0<6$], therefore, the graph of $2x + 3y \geq 6$ is that half of XOY-plane which lies above the line AB (Including the points of the line AB). The required graph is shown shaded in fig.

Thus, solving the inequation $ax + by \leq c$ by graphical method involves the following steps:

Step I. Consider the equation $ax + by = c$ and plot the resulting line. In case of strict inequalities < or >, draw the line dotted, otherwise mark it thick.

Step II. Choose a point [If possible (0, 0)], not lying on this line. Substitute its coordinates in the inequation. If the inequation is satisfied, then shade the portion of the plane which contains the chosen points; otherwise shade the portion which does not contain this point.

The shaded portion represents the solution set. The dotted line is not a part of shaded region while thick line is a part of it.

Simultaneous Inequations: The solution set of a system of linear inequations in two variables is the set of all points (x, y) which satisfy all the inequations in the system simultaneously. So, we find the region of the plane, common to all the portions comprising the solution sets of given inequations. When there is no region common to all the solution of the given inequations, we say that the solution set is empty.

The linear inequations are also known as linear constraints.

Illustration: Draw the diagram of the solution set of the inequations $2x + 3y \geq 6$, $x + 4y \leq 4$, $x \geq 0$ and $y \geq 0$.

Solution: Consider the equations,

$2x + 3y = 6$, $x + 4y = 4$, $x = 0$ and $y = 0$.

Now $2x + 3y = 6 \Rightarrow \frac{x}{3} + \frac{y}{2} = 1.$

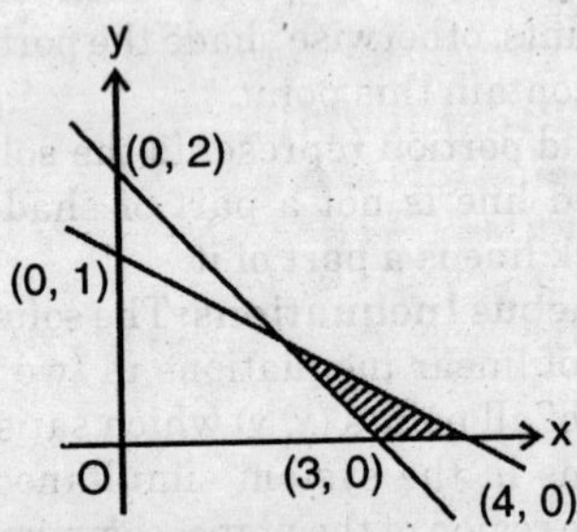

This meets the axes at (3, 0) and (0, 2). Join these points and draw a thick line. Clearly, the portion not containing (0, 0) represents the solution set of the inequation, $2x + 3y \geq 6$.

Agian $x + 4y = 4 \Rightarrow \frac{x}{4} + \frac{y}{1} = 1.$

This line meets the axes at (4, 0) and (0, 1). Join these points and draw a thick line. Clearly, the portion containing (0, 0) represents the solution set of the inequation $x + 4y \leq 4$

Clearly, $x \geq 0$ is represented by y-axis and the portion on its right hand side.

Also $y \geq 0$ is represented by x-axis and the portion above x-axis.

Hence, the shaded region represents the solution set of the given inequation.

Linear Programming: The idea of linear programming was first introduced by a Russian mathematician L.V. Kantorvich. In 1947, George B. Dantzig developed a superior technique of computation popularly known as the "simplex method". He developed it for the purpose of scheduling highly complex procurement activities of the United Sates Air Force. The development of electronic computers has made a significant role in the growth of linear programming because computers can easily and quickly solve complicated problems which may not be solved otherwise.

Meaning: Programming means systematic planning or decision making. It is a technique for solving optimization (maximization or minimization) problems subjects to certain constrain. Out of all permissible allocations of resources, we have to decide one which will minimize the total cost or maximize the total profits. Linear Programming is a device which is used in decision making in business for obtaining optimum values of quantities subject to certain constraints when the relationship involved in the problem are linear.

Methods of Solving linear Programming Problems: There are two methods of solving a linear programming problem.

1. Graphical Method and
2. Simplex method

The simplex method is beyond the scope of this book. We shall explain the graphical method of solving a linear programming problem. For this we need the following back ground.

(*i*) A set S of points in a plane is said to be convex, if the line segment joining any two points in it, lies in it completely, *i.e.*, if we take any two points A and B in the set S, then the segment $[AB]$ lies in S. A circle, a square, a rhombus, a polygon are examples of convex set. In fig. (*i*) and (*iii*) are convex sets, where as (*ii*) is not a convex set.

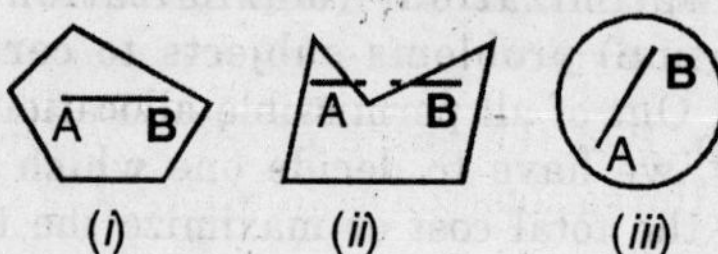

(*ii*) If we have a system of linear inequation in two variables, then the set of points (x, y) for which all the inequations of the system hold true is either empty, or convex region bounded by straight lines (a Convex Polygon) or an unbounded region with straight line boundaries.

(*iii*) The set of points, whose co-ordinates satisfy the constraints of a linear programming problem, is said to be the feasible region.

(*iv*) The optimum value (maximum or minimum) of the objective function is obtained at a vertex of the feasible region. If there are more than one point (vertices) where the objective function is optimum (max, or min.), then every point on the line segment joining any two such vertices optimizes the objective function.

Thus, solving a linear programming problem by the graphical method involves the following steps.

First step. Plot the graph of the inequalities describing the various constraints (structural) on the graph paper.

Second step. Find the portion of the graph paper in the first quadrant ($\because$ of non-negativity constraints) which is common to the graphs plotted in the first step. Locate the extreme points (*i.e.*, corner points) of this region (known as feasible region).

Third step. Find the value of the objective function corresponding to each corner point. The points which corresponds to the optimum (*i.e.*, max. or min.) value of the objective function is (are) required solution (s) of the given linear programming problems.

Remark. Though (0, 0) may be a corner point of the feasible region in some problem, but it is not to be examined for the optimum solution.

Illustration: Find the maximum and minimum values of $5x + 2y$ subject to constraints

$-2x - 3y \le -6$ *i.e.*, $2x + 3y \ge 6$

$x - 2y \le 2$

$6x + 4y \le 24$

$-3x + 2y \le 3$

$x \ge 0$ and $y \ge 0$

Solution: The bounding lines for feasible region are

$$2x + 3y = 6 \Rightarrow \frac{x}{3} + \frac{y}{2} = 1 \quad ...(1)$$

$$x - 2y = 2 \Rightarrow \frac{x}{2} + \frac{y}{-1} = 1 \quad ...(2)$$

$$6x + 4y = 24 \Rightarrow \frac{x}{4} + \frac{y}{6} = 1 \quad ...(3)$$

$$-3x + 2y = 3 \Rightarrow \frac{x}{-1} + \frac{y}{\frac{3}{2}} = 1 \quad ...(4)$$

and $x = 0, y = 0$ ...(5)

Clearly the line (1) meets the axes at (3, 0) and (0, 2) the line (2) meets the axes at (2, 0) and (0, – 1) the line (3) meets the axes at (4, 0) and (0, 6) the line (4) meets the axes at (–1, 0) and $\left(0, \frac{3}{2}\right)$. Also $x = 0$ is the y-axis and $y = 0$ is the x-axis.

Plotting the above points and joining them we get the lines l_1, l_2, l_3 and l_4 as shown in fig. The feasible region is the interior of the quadrilateral *ABCD* as shown in fig. Solving equations (1) and (2) we find that the vertex *A* is $\left(\frac{18}{7}, \frac{2}{7}\right)$.

Similarly the Coordinates of vertices B, C and D are $\left(\frac{7}{2},\frac{3}{4}\right)$, $\left(\frac{3}{2},\frac{15}{4}\right)$ and $\left(\frac{3}{13},\frac{24}{13}\right)$.

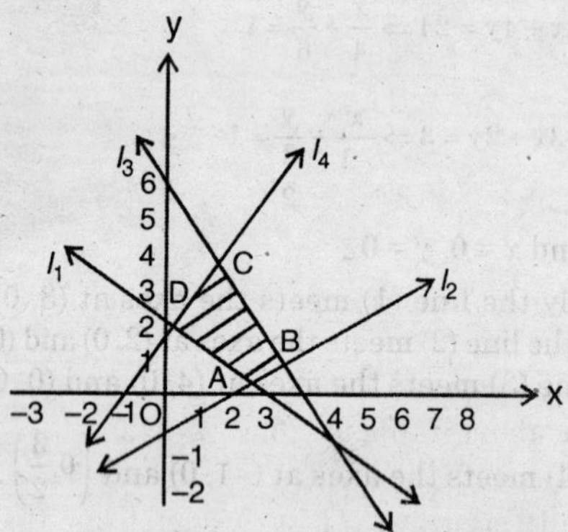

The Cooresponding values of objective function $5x + 2y$ are tabulated below:

Vertex	Co-ordinates	Values of $5x = 2y$
A	$\left(\frac{18}{7}, \frac{2}{7}\right)$	$5 \times \frac{18}{7} + 2 \times \frac{2}{7} = \frac{94}{7} = 13\frac{3}{7}$
B	$\left(\frac{7}{2}, \frac{3}{4}\right)$	$5 \times \frac{7}{2} + 2 \times \frac{3}{4} = \frac{38}{2} = 19$
C	$\left(\frac{3}{2}, \frac{15}{4}\right)$	$5 \times \frac{3}{2} + 2 \times \frac{15}{4} = 15$
D	$\left(\frac{3}{13}, \frac{24}{13}\right)$	$5 \times \frac{3}{13} + 2 \times \frac{24}{13} = \frac{63}{13} = 4\frac{11}{13}$

From above table we observe that maximum value 19 occurs at $B\left(\frac{7}{2}, \frac{3}{4}\right)$ and minimum value occurs $4\frac{11}{13}$ at $D\left(\frac{3}{13}, \frac{24}{13}\right)$.

18

PROBABILITY

Def: A mathematically measure of uncertainty is known as probability.

Experiment: An experiment in any action which we do or intend to do. An experiment is performed for achieving some objective. For example, coin is tossed in order to choose batting or fielding first in cricket match.

Random Experiment: Word "random experiment" signifies that the experiment is performed in a haphazard way. It is performed in an unplanned manner. The word random signifies that the possibility of a particular outcome as a result of performing an experiment is not a certainty.

Tossing of a coin and throwing a die are examples of random experiment.

Outcome: The result of a random experiment is called an outcome.

Sample space: The set of all possible outcomes of a random experiment is known as sample space. For example

(*i*) The sample space in a throw of a die is the set {1, 2, 3, 4, 5, 6}

(*ii*) If we toss coins, the sample space in this consists of four outcomes.

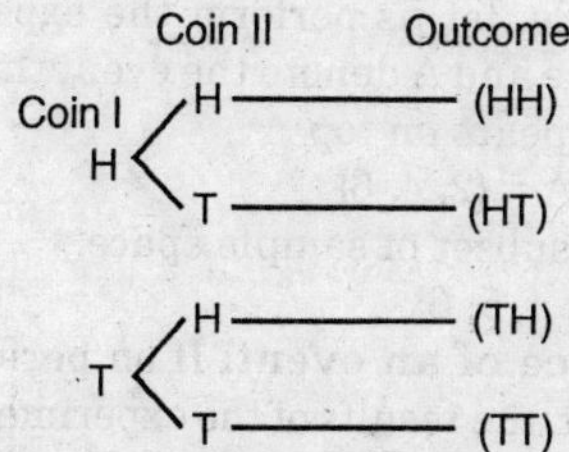

The sample space is the set {(H, H), (H, T), (T, H), (T, T)}

Trial: The performance of a random experimental is called a trial.

Events: An event is a set of experimental outcomes, or in other words it is a subset of sample space S.

Note:

(*i*) an event may contain any number of the elements of sample space S.

(*ii*) Elements of a sample space are called elementary events.

(*iii*) As yet S is a subset of S therefore S is also an event.

(*iv*) We know ϕ is subset of every set so, ϕ is also an example of an event and sometimes is called impossible event.

For example, let us perform the experiment of tossing a die and A denote the event that an even number appears on top

Obviously A = {2, 4, 6}

which is a subset of sample space.

S = {1, 2, 3, 4, 5, 6}.

Occurrence of an event: If on performing an experiment, the results of the experiment belongs to the subset defining an event then the event is said to have occurred. Event is said to have not occurred if the result of the experiment does not belong to the subset defining the event.

Simple event: An event consisting of only one sample point of sample space is called a simple event.

An event consisting of more than one sample point is called a compound event.

Mutually exclusive events: The events are said to be mutually exclusive or incomplete if no two or more of them can happen simultaneously in the same trial. For example, appearing of head and tail in tossing of a coin are mutually exclusive as there is only one possibility – either head or tail. Both head and tail cannot occur.

Equally likely events: The events are said to be equally likely if there is no reason except any one in preference to any other. For example, in tossing of a coin, appearing a head or tail are equally likely.

Exhaustive events: All possible outcomes of an event are known as exhaustive events. For example, in a throw of a single dice the exhaustive events are six i.e. 1, 2, 3, 4, 5, 6. If two dice are thrown the exhaustive events would be $6 \times 6 = 36$.

Favourable events: The cases which ensures the occurrence of the event are callled favourable. For example, when two dice are thrown the number of cases favourable for getting a sum 6 is 5 i.e. (1, 5) , (5, 1), (2, 4) (4, 2) and (3, 3).

Independent events: Two or more events are said to be independent if the happening or non-happening or any other event. For example the result of first toss of a coin does not influence the result of second toss of a coin.

Dependent events: If the happening of an event influences the happening of the other event, the events are said to be dependent events.

For example, if a card is drawn from a pack of shuffled cards and not replaced before drawing the second card, then the second draw is dependent on the first one.

Algebra of events: In a single throw of a dice, let A denote the event "The appearance of the even number" and B denotes the event "The appearance of the number of greater than 3".

$\therefore$ Set A = {2, 4, 6}

Set B = {4, 5, 6} and S = {1, 2, 3, 4, 5, 6}

(*i*) Consider a new event "A or B" which occurs when A or B or both occur. Clearly the event "A or B" will be represented by

$A \cup B = \{2, 4, 5, 6\}$

(*ii*) Consider another event "A and B" which can occur only when both occurs. Clearly "A and B" will be represented by

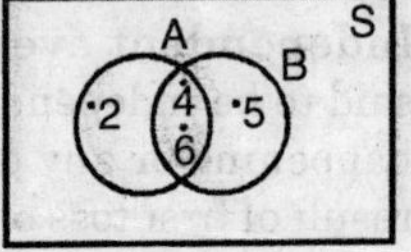

$A \cap B = \{4, 6\}$

Complementary events: Event which is complementary to the event E is set of those elements of sample space which do not belong to E. Event complement to E is denoted by E^c or $\overline{E}$. It is also called negation of E.

Probability of an event: If a trial results in 'n' exhaustive, mutually exclusive and equally likely cases, out of which 'm' are favourable in the occurrence of an event A, then the probability of occurrence of event A, usually denoted P (A) is given by

$$P(A) = \frac{\text{Number of favourable cases}}{\text{Exhaustive number of cases}} = \frac{m}{n}.$$

Note:

1. If the probability of an event A is zero then we say that the event A is impossible, and if the probability of an event is one then we say that the event A is sure or certain.
2. If the probability of an event A is 'p', $0 \le p \le 1$ *i.e.*, $P(A) = p$, then we say
   - (*i*) The chance that event A happen is p
   - (*ii*) The chance that event A fails to happen is $1 - p$
   - (*iii*) Odds in favour of event A are $p : (1 - p)$
   - (iv) Odds against the event A are $(1 - p) : p$

3. If an event A can happen in '*m*' ways and fail in '*n*' ways, then the following three statement are equivalent.

(*i*) $P(A) = \frac{m}{m+n}$

(*ii*) The odds in favour of the event A are

$$\frac{m}{m+n}; \left(1 - \frac{m}{m+n}\right) \text{ i.e., } m : n$$

(*iii*) The odds against the event A are

$$\left(1 - \frac{m}{m+n}\right); \frac{m}{m+n} \text{ i.e., } n : m$$

(*iv*) Probability of non-happening of event A is denoted by

$$P(\overline{A}) = \frac{\text{Number of favourable cases (i.e. unfavourable)}}{\text{Exhaustive number of outcomes}}$$

$$= \frac{n-m}{m} = 1 - \frac{m}{n} = 1 - P(A)$$

$$\Rightarrow P(A) + P(\overline{A}) = 1$$

Theorems on Probability

If A and B are two mutually exclusive events of a random experiment, the probability of occurrence of the event "A or B" is the sum of the probabilities of the events, A and B, or

$$P\,(A \text{ or } B) = P\,(A) + P(B)$$

if A and B are mutually exclusive events or

In the language of subsets of the sample space S, theorem is written as follows:

If $A \cap B = \phi$ then $P\,(A \cup B) = P\,(A) + P\,(B)$

Cor I. If $A_1, A_2, A_3 \ldots A_k$ are mutually exclusive events, then

$P(A_1 \cup A_2 \cup A_3 \ldots \cup A_k) = P\,(A_1) + (A_2) + \ldots + P(A_k)$

II. If A and B are two events associated with a random experiment then,

$P(A \text{ or } B) = P(A) + P(B) - P\,(A \text{ and } B)$ or

In the language of the theory of sets it can be written as

$P\,(A \cup B) = P(A) + P(B) - P\,(A \cap B)$

III. For every event A associated with random experiment, we have

$P\,(\text{not } A) = 1 - P\,(A)$

or $\quad P(\overline{A}) = 1 - P\,(A)$

IV. If the event A implies the event, B, then

$$P\,(A) \le P\,(B)$$

V. For any two events A and B

$$P(\overline{A} \cap B) = P\,(B) - P\,(A \cap B)$$

VI. If $B \subset A$, then

$$P(A \cap \overline{B}) = P(A) - P(B)$$

VII. If A and B are any two events (subsets of sample space S) and are disjoint, then

$$P(A \cup B) = P(A) + P(B) - P(A \cap B)$$

Conditional Probability: Let A and B be two events associated with sample space S. Then the probability of occurrence of the event A when B has occurred is called the conditional probability of A and is denoted by P (A/B)

$\therefore$

$$P(A / B) = \frac{\text{Number of cases favourable to both A and B}}{\text{Numbe of cases in which B happens}}$$

$$= = \frac{n(A \cap B)}{n(B)}$$

Multiplicative Rule of Probability or Theorem of Compound Probability: The probability of the simultaneous occurrences of two dependent events is given by the product of the unconditional probability of one of them multiplied by the conditional probability of the other asssuming that the first has already occurred. Symbolically for two events

$$P(A \cap B) = P(B/A).\ P(A),\ P(A) \neq 0$$

$$P(A \cap B) = P(A/B).\ P(B),\ P(B) \neq 0$$

Multiplicative Rule for Independent Events: The probabilityof simultaneous occurrences of two independent events is equal to the product of their individual probabilities *i.e.*, for two independent events A and B,

$P(A \cap B) = P(A).\ P(B)$; it can also be written as

$P(AB) = P(A)\ .\ P(B)$

Baye's Theorem: If $E_1, E_2, E_3 \ldots E_n$ form a set of mutually exclusive and exhaustive events and A is any other event, then

$$P(E_i / A) = \frac{P(E_i)P(A / E_i)}{P(E_1)P(A / E_1) + P(E_2)P(A / E_2) \ldots + P(E_n)P(A / E_n)}$$

Random Variable: A random variable is a real valued function defined over the sample space of an experiment i.e., a variable whose value is a number determined by the sample point (outcome of the experiment) of a sample space is called a random variable. A random variable is denoted by X, Y, Z ... etc.

For example, let X be a random variable which is the number of heads obtained in two independent tosses of an unbiased coin.

S = {HH, HT, TH, TT}

Then X (HH) = 2, X (HT) = 1, X (TH) = 1, X (TT) = 0

$\therefore$ X can take values 0, 1, 2

Mean of Random Variable:

Mean $\mu = \frac{\Sigma p_i x_i}{\Sigma p_i}$

Variance of Random Variable:

$\sigma^2 = \Sigma p_i x_i^2 - (p_i x_i)^2$

Discrete and Continuous Random Variables: If the random variable X assumes only a finite or countably infinite set of values it is called *discrete random variable.* For example, marks obtained by students in an examination, the number of students in a school, the number of effective bulbs etc., are all discrete random variable.

But if the random variable X can assume infinite and uncountable set of values it is said to be *a continuous random variable.* For example, the age, height or weight of a group of students are all continuous random variable.

Binomial Distribution: Binomial distribution is a probability distribution which is obtained when the probability *p* of the happening of an event is same in all the trials. For example the probability of getting head, when the coin is tossed a number of times must remain same in each toss *i.e.* $\frac{1}{2}$.

$P(r) = {}^nC_r p^r q^{n-r}$

Here, n is the trials, p is the probability of success and q be the probability of failure.

Mean of binomial distribution. $(\mu) = np$

Variance of binomial distribution $= (\sigma^2) = npq$

Properties of binomial distribution:

1. The binomial distribution has its

   Mean $= np$, Variance $= npq$ or S.D. $= \sqrt{npq}$

   $\mu_1' = np, \mu_2' = n^2p^2 - np^2 + np, \mu_2 = \mu_2' - (\mu_1')^2$.

2. Binomial distribution is symmetrical if

   $p = q = \frac{1}{2}$

3. The shape and location of a Binomial distribution changes as p changes for a given n or as n changes for a given p.
4. Binomial distribution can be represented graphically.

19

VECTOR ALGEBRA

Scalar quantity: Quantities which have magnitude but no direction are called scalar quantities or simply scalars. For example, mass, speed, time etc.

Vector quantities: Quantities which have magnitude as well as direction are called vector quantities or simply vectors. For example, force, velocity etc.

Directed lines segment: A line segment (portion of line) which is assigned a definite direction by specifying its initial and terminal points is called a directed line segment. The directed lined segment AB has

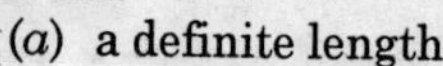

(*a*) a definite length

(*b*) a definite direction, from A to B, indicated by the arrow head.

Representation of a vector: A directed line segment represents a vector. A vector whose magnitude is proportional to the length AB and whose direction is from A to B and is denoted by $\overrightarrow{AB}$ where A is called the initial point and B is the terminal point of the vector $\overrightarrow{AB}$. The magnitude of the vector $\overrightarrow{AB}$ is denoted by $\left|\overrightarrow{AB}\right|$ which is read as modulus of $\overrightarrow{AB}$ or simply $\overrightarrow{AB}$. $\left|\overrightarrow{AB}\right| = AB$.

If the directed line segment AB is a part of the line AB, then the line l is called the support of vector $\overrightarrow{AB}$.

Type of vectors:

(*i*) **Like and unlike vectors:** Two (or more than two) vectors are said to be like vectors if they have the same direction (no matter what their magnitudes are). Unlikely vector have opposite directions.

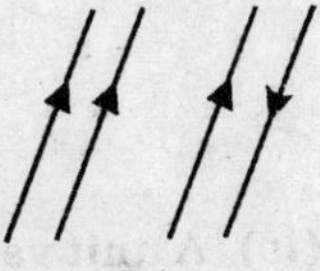

(*ii*) Null vector or zero vector : A vector whose magnitude is 0 is called a null or zero vector and is represented by $\vec{O}$. The initial and terminal points of a zero vectors are co-incident and its direction is arbitrary. Thus,

$$\vec{AA} = \vec{BB} = \ldots\ldots = \vec{O} \text{ and } \left|\vec{O}\right| = 0.$$

A non-zero vector is called proper vector.

(*iii*) Equal vectors : Two vectors are said to be equal if they have same magnitude and same direction. Thus, $\vec{AB} = \vec{CD}$

if (*a*) $|\vec{AB}| = |\vec{CD}|$ *i.e.* AB = CD

(*b*) $\vec{AB}$ and $\vec{CD}$ are like vectors.

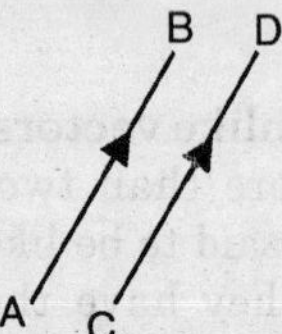

(*iv*) A unit vector : A vector whose magnitude is unity (one) is called a unit vector. If $\vec{a}$ is a unit vector then $|\vec{a}| = a = 1$. The unit vector in the direction of $\vec{a}$ is denoted by $\hat{a}$ and read as a-cap. The unit vectors in the positive directions of the axes of x, y and z are denoted by $\hat{i}, \hat{j}, \hat{k}$.

(v) **Coplaner Vectors:** Two or more vectors are said to be coplaner if they are parallel to the same plane.

(*vi*) **Collinear vectors :** Two or more vectors are said to be collinear if they are parallel to the same line irrespective of their magnitude.

(*vii*) **Negative vectors :** Two or more vectors are called negatives of each other if they have same magnitude but opposite direction. Negative of $\vec{a}$ is denoted by $-\vec{a}$.

(*viii*) **Localised vectors :** A vector having a fixed initial points is called a localised vectors.

(*ix*) **Co-initial vector :** Two or more vectors are said to be co-initial vectors if they have the same initial point. For example $\overrightarrow{AB}$, $\overrightarrow{AC}$, $\overrightarrow{AD}$ are co-initial vectors with initial point A.

(x) Free vectors : Vectors whose directions and magnitudes are known but the initial point and the support are not known are called free vectors.

Addition of two vectors $\vec{a}$ and $\vec{b}$

(a) *Sum of two vectors by: Triangle Law*

$$\overrightarrow{AB} + \overrightarrow{BC} = \overrightarrow{AC}$$

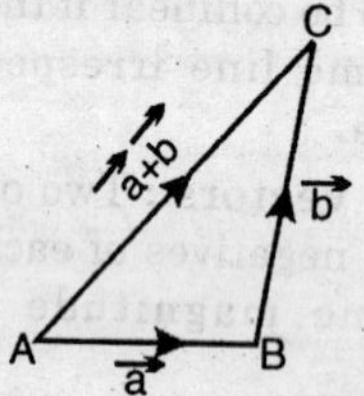

In words : If two vectors are represented by the two sides of a triangle taken in order, then their sum is represented by the third side of the triangle taken in opposite direction.

Note : Two vectors can be added by triangle law only if the terminal points of the directed segment representing one vector is the same as the initial point of the directed segment representing the second vector.

(*b*) *Sum of two vectors by parallelogram*:

$\vec{AB} + \vec{AC} = \vec{AD}$

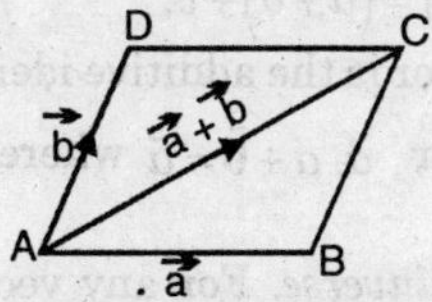

Note : These two methods of addition namely triangle and parallelogram law are identical.

(*c*) *Difference of two vectors :* The difference of two vectors $\vec{a}$ and $\vec{b}$ is defined as the addition of one to the negative of the other.

Thus, $\vec{a} - \vec{b} = \vec{a} + \left(-\vec{b}\right)$.

Properties of addition of vectors

1. Vector addition is cumulative i.e. if $\vec{a}$ and $\vec{b}$ are two vectors, then $\vec{a}+\vec{b}=\vec{b}+\vec{a}$.
2. Vector addition is associative i.e. for any three vectors $\vec{a}, \vec{b}, \vec{c}$.
$\vec{a}+(\vec{b}+\vec{c})=(\vec{a}+\vec{b})+\vec{c}$.
3. Zero vector is the additive identity i.e from any vector $\vec{a}, \vec{a}+\vec{0}=\vec{a}$ where $\vec{0}$ is the null vector.
4. *Additive inverse.* For any vector $\vec{a}$, there exists the vector $-\vec{a}$ such that $\vec{a}+(-\vec{a})=0$.

Properties of multiplication of vector by a scalar

1. *Associative law.* If $\vec{a}$ any vector and m, n are any scalar, then
$m(n\vec{a})=(mn)\vec{a}$
2. *Distributive law.* If $\vec{a}$ is any vector and k, l are any scalars, then
$(k+l)\vec{a}=k\vec{a}+l\vec{a}$.
3. *Distributive law.* If $\vec{a}$ and $\vec{b}$ are any two vectors and k is a scalar, then

$$k(\vec{a}+\vec{b}) = k\vec{a} + k\vec{b}$$

Section formula : If $\vec{a}$ and $\vec{b}$ are the position vectors of two points A and B then the point C which divides AB in the ratio of m ; n where m and n are the positive real numbers has the position vector.

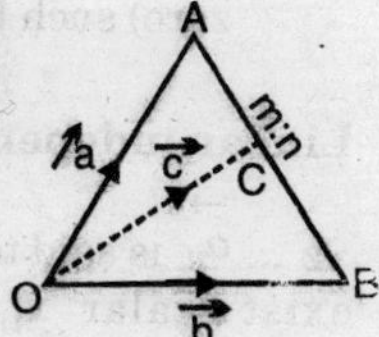

$$\vec{c} = \frac{n\vec{a} + m\vec{b}}{m+n}$$

Cor. Mid-point formula. If $m = n$, then C the mid-point of AB, $\vec{c} = \dfrac{\vec{a}+\vec{b}}{2}$

Linear combination :

(*i*) A vector $\vec{r}$ is said to be a linear combination of the vectors $\vec{a}, \vec{b}, \vec{c} \ldots$ if there exists scalars x, y, z ... such that $\vec{r} = x\vec{a} + y\vec{b} + z\vec{c} + \ldots$

(*ii*) **Linear dependent:** A system of vectors, $\vec{a_1}, \vec{a_2} \ldots \vec{a_n}$ is said to be linearly dependent if there exists scalars $x_1, x_2 \dots x_n$ (not all zero) such that $x_1 \vec{a_1}, x_2 \vec{a_2} + \dots x_n \vec{a_n} = 0$

Linear independent: A system of vectors $\vec{a_1}$, $\vec{a_2} \dots \vec{a_n}$ is said to be linearly dependent if there exist scalar $x_1, x_2 \dots x_n$ (all zero) such that

$x_1 \vec{a_1} + x_2 \vec{a_2} + \dots x_n \vec{a_n} = 0.$

Remark: $\vec{r}$ and $x\vec{r}$ are collinear, where x is a scalar, $x\vec{a} + y\vec{b}$ represents a vector coplanar with vectors $\vec{a}$ and $\vec{b}$ where x, y are scalars.

Theorenı1. If $\vec{a}, \vec{b}$ be two non-zero, non-collinear vectors and x, y are two scalars such that $x\vec{a} + y\vec{b} = 0$, then $x = 0, y = 0$.

Theorem 2. If $\vec{a}, \vec{b}, \vec{c}$ be three non-zero, non-coplanar vectors and x, y, z are three scalars such that, $x\vec{a} + y\vec{b} + z\vec{c} = 0$, then $x = 0, y = 0, z = 0$.

Theorem 3. *Resolution of a vector in terms of coplanar vectors.* If $\vec{a}, \vec{b}$ be two given non-collinear vectors, then every vector $\vec{r}$ can be expressed uniquely as a linear combination $x\vec{a} + y\vec{b}$; x, y being scalars.

Theorem 4. *Non-coplanar vector*

If $\vec{a}$, $\vec{b}$, $\vec{c}$ be three given non-coplanar vectors, then any vector $\vec{r}$ can be expressed uniquely as a linear combination, $x\vec{a} + y\vec{b} + z\vec{c}$, x, y, z being scalars.

Unit vector is that vector whose magnitude is unity we denote unit vectors along OX, OY, OZ by $\vec{i}, \vec{j}, \vec{k}$ respectively.

Product of two vectors: The product of two vectors is defined in two ways: (*i*) Scalar product (*ii*) Vector product.

(*i*) Scalar product of two vectors is always a scalar quantity.

(*ii*) Vector product of two vectors is always a vector quantity.

Scalar product: The scalar product of two vectors $\vec{a}$ and $\vec{b}$ with magnitude a and b respectively is defined by the real number $|\vec{a}|\ |\vec{b}| \cos\theta$, where θ is the angle between the direction of a and b. Here, θ is restricted to the interval $0 \le \theta \le \pi$.

It makes no difference whether θ or –θ is chosen as $\cos\theta = \cos(-\theta)$.

Notation: The scalar product of two vectors is written as $\vec{a}\ \vec{b}$ or $\left(\vec{a}.\ \vec{b}\right)$, $\vec{a}\ .\ \vec{b}$ is read as $\vec{a}$ dot $\vec{b}$. Because of this the scalar product is sometimes called the dot product.

Aid to memory: $\vec{a}.\ \vec{b} = |\vec{a}|\ |\vec{b}| \cos\theta$, which is +ve, –ve or zero according as θ is acute, obtuse or a right angle.

Geometrical interpretation:

The scalar product of two vector is the product of the modulus of either vector and the scalar component of the other in its direction.

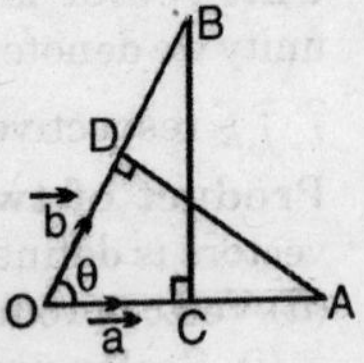

Aid to memory: If $\vec{a}$ and $\vec{b}$ are two vectors, then the projection of b on the direction of $\vec{a} = \dfrac{\vec{a}.\ \vec{b}}{|\vec{a}|}$.

Condition of perpendicularity:

Theorem. If $\vec{a}$ and $\vec{b} = 0$, are perpendicular vectors, then $\vec{a}.\ \vec{b} = 0$

Conversely: If $\vec{a}.\vec{b}=0$ then either at least one of the two vectors is a zero vector or two vectors are perpendicular.

Some important results:

1. *When the two vectors are like parallel,*

   Here $\theta = 0° \therefore \cos\theta = \cos 0° = 1$.

   $\therefore \vec{a}.\vec{b} = |\vec{a}|.|\vec{b}| \cos\theta = ab\,(1) = ab.$

2. When the two vectors unlike parallel,

   Here $\theta = \pi$

   $\therefore \cos\theta = \cos\pi = -1$

   $\therefore \vec{a}.\vec{b} = |\vec{a}|.|\vec{b}| \cos\theta = ab\,(-1) = -ab$

3. $\vec{a}.\vec{a} = a^2.$

4. Since $\vec{i}, \vec{j}, \vec{k}$ are vectors perpendicular to each other,

   $\therefore \vec{i}.\vec{j} = \vec{j}.\vec{k} = \vec{k}.\vec{i} = 0$

   and $\vec{i}^2 = \vec{j}^2 = \vec{k}^2 = 1$

Aid to memory: These results can be remembered with the help of the following table.

•	$\vec{i}$	$\vec{j}$	$\vec{k}$
i	1	0	0
j	0	1	0
k	0	0	1

Properties of scalar or dot products

Property 1. Comulative law. If $\vec{a}, \vec{b}$ be any two vectors, then $\vec{a}.\vec{b} = \vec{b}\,.\,\vec{a}$.

Property 2. Associative law does not hold. If $\vec{a}, \vec{b}, \vec{c}$ be any three vectors, then $\left(\vec{a}.\,\vec{b}\right).\vec{c} \neq \vec{a}.\left(\vec{b}.\vec{c}\right)$.

Property 3. If x is any scalar then

$$(x\,\vec{a}).\,\vec{b} = x\left(\vec{a}\,.\,\vec{b}\right) = \vec{a}\,.\left(x\vec{b}\right).$$

Property 4. Distributive law. If $\vec{a}.\,\vec{b}, \vec{c}$ are any three vectors, then

$$\vec{a}\,.\left(\vec{b}+\vec{c}\right) = \vec{a}\,.\,\vec{b} + \vec{a}\,.\,\vec{c}$$

Cor. $\vec{a}\,.\left(\vec{b}.-\vec{c}\right) = \vec{a}\,.\left[\vec{b}+(-\vec{c})\right] = \vec{a}\,.\,\vec{b} + \vec{a}\,.\,(-\vec{c})$

$= \vec{a}\,.\,\vec{b}\,.-\vec{a}\,.\,\vec{c}$

Property 5. $\vec{a}\,.\,\vec{a} \geq 0$ where a is any vector.

Theorem. If $a = (a_1, a_2, a_3)$ and $b = (b_1, b_2, b_3)$, then $(\vec{a} \cdot \vec{b}) = (a_1b_1, a_2b_2, a_3b_3)$.

Angle between two vectors: If θ be the angle between two vectors $\vec{a} = (a_1, a_2, a_3)$ and $\vec{b} = (b_1, b_2, b_3)$ then

$$\cos\theta = \frac{a_1b_1 + a_2b_2 + a_3b_3}{\sqrt{a_1^2 + a_2^2 + a_3^2}\sqrt{b_1^2 + b_2^2 + b_3^2}}$$

Cor. *Condition of parallelism*

$$\frac{a_1}{b_1} = \frac{a_2}{a_2} = \frac{a_3}{b_3}.$$

Cor. 2 *Condition of perpendicularity*

$$a_1b_1 + a_2b_2 + a_3b_3 = 0.$$

Work done by a force : A force acting on a particle is said to do work when the particle is displaced in a direction which is not perpendicular to the direction of the force. The work done is scalar quantity and its measure is defined to the product of force and displacement. Thus $\vec{F}$, $\vec{d}$ be vectors representing the force and the displacement respectively inclined at an angle θ, the measure of the work done is $\vec{F}d \cos\theta = \vec{F}.\vec{d}$.

Note: Work done is zero if $\vec{d}$ is perpendicular to $\vec{F}$ because in this case, $\cos\theta = \cos\frac{\pi}{2} = 0$.

Vector product: The vector product of two vectors $\vec{a}$ and $\vec{b}$ is a vector whose magnitude is $|\vec{a}|\ |\vec{b}|$ $\sin\theta$, where θ is the angle between two direction is that of a unit $\hat{\vec{n}}$ perpendicular to both $\vec{a}$ and $\vec{b}$.

Notation: The vector product of two vectors is written as $\vec{a}\times\vec{b}$ or $[\vec{a}\times\vec{b}]$

$\vec{a}\times\vec{b}$ is read as $\vec{a}$ cross $\vec{b}$, because this vector product is sometimes called the the cross product.

Remember: $\vec{a}\times\vec{b} = |\vec{a}|\ |\vec{b}| \sin\theta\, \vec{n}$ where $\vec{a}, \vec{b}, \hat{\vec{n}}$ form a right handed system.

Some important results

1. *When two vectors are parallel,*
   Here $\theta = 0°$.
   $\therefore \sin\theta = 0$

$\vec{a} \times \vec{b}, = |\vec{a}|\ |\vec{b}| \sin\theta\ \hat{\vec{n}} = \vec{0}$

Particular case. $\vec{a} \times \vec{a} = \vec{0},\ \vec{b} \times \vec{b} = \vec{0}.$

Cor. When $\vec{a} \times \vec{b} = 0$ then either $\vec{a} = \vec{0}$ or $\vec{b} = \vec{0}$ or $\vec{a}$ and $\vec{b}$ are parallel vectors.

2. *When two vectors are perpendicular.*
   Here $\theta = 90°$ $\quad \therefore \sin\theta = 1$

$$\therefore\ \vec{a} \times \vec{b} = |\vec{a}|\ |\vec{b}| \sin\theta\ \hat{\vec{n}} = |\vec{a}|\ |\vec{b}|\ \hat{\vec{n}}$$

3. (*a*) *When $\vec{a}$ and $\vec{b}$ are unit vectors.*

   Here $\vec{a} \times \vec{b} = (1) \sin\theta\ \hat{\vec{n}}$

$$\therefore\ |\vec{a} \times \vec{b}| = \sin\theta.$$

   (*b*) When $\vec{a}$ *and* $\vec{b}$ *are not unit vectors.* Here

$$\vec{a} \times \vec{b} = |\vec{a}|\ |\vec{b}| \sin\theta\ \hat{\vec{n}}$$

$$\therefore\ |\vec{a}\ \vec{b}| = |\vec{a}|\ |\vec{b}| \sin\theta \Rightarrow \sin\theta = \frac{|\vec{a} \times \vec{b}|}{|\vec{a}|\ |\vec{b}|}$$

4. *Vector products of unit vectors.* $\vec{i}, \vec{j}, \vec{k}$. Since $\vec{i}, \vec{j}, \vec{k}$ form a right handed system of mutually perpendicular vectors.

$\therefore \quad \vec{i} \times \vec{j}$ is a vector having modulus unity and direction parallel to k.

$$\therefore \quad \vec{i} \times \vec{j} = \vec{k} = -\vec{j} \times \vec{i}$$

$$\vec{i} \times \vec{k} = \vec{i} = -\vec{k} \times \vec{j}$$

$$\vec{k} \times \vec{i} = \vec{j} = -\vec{i} \times \vec{k}$$

$$\vec{i} \times \vec{i} = \vec{j} \times \vec{j} = \vec{k} \times \vec{k} = 0$$

These results can be remembered with the help of following table

$\times$	$\vec{i}$	$\vec{j}$	$\vec{k}$
$\vec{i}$	$\vec{0}$	$\vec{k}$	$-\vec{j}$
$\vec{j}$	$-\vec{k}$	$\vec{0}$	$\vec{i}$
$\vec{k}$	$\vec{j}$	$-\vec{i}$	$\vec{0}$

Properties of vector or cross product

Property 1. *Commulative law.* It does not hold

If $\vec{a}, \vec{b}$ are vectors, then $\vec{a} \times \vec{b} \neq \vec{b} \times \vec{a}$

Another form $\vec{a} \times \vec{b} = -\vec{b} \times \vec{a}$

roperty 2.

(a) *Vector product is associative w.r.t. a scalar*

i.e. $(m\vec{a})\times\vec{b} = \vec{m}(\vec{a}\times\vec{b}) = \vec{a}\times(m\,\vec{b})$, *where m is any scalar.*

(b) *Vector product is not associative w.r.t. to vector.*

i.e. $\vec{a}\times(\vec{b}\times\vec{c}) \neq (\vec{a}\times\vec{b})\times\vec{c}$.

roperty 1. *Vectro product in terms of rectangular omponents of vectors*

$$\vec{a} = a_1\,\vec{i} + a_2\,\vec{j} + a_3\,\vec{k}$$

$$\vec{b} = b_1\,\vec{i} + b_2\,\vec{j} + b_3\,\vec{k}$$

$$\vec{a}\times\vec{b} = \begin{bmatrix} \vec{i} & \vec{j} & \vec{k} \\ a_1 & a_2 & a_3 \\ b_1 & b_2 & b_3 \end{bmatrix}$$

roperty 4. Distributive law. If $\vec{a}, \vec{b}, \vec{c}$ are three ectors then

$$\vec{a}\times(\vec{b}+\vec{c}) = \vec{a}\times\vec{b} + \vec{a}\times\vec{c}$$

---

(20)

FUNCTIONS, LIMITS AND CONTINUITY

Function: Let X = {1, 2, 3} and Y = {4, 5, 6, 7}

Let us consider the following association between the sets X and Y. Let us associate 1 to 4; 2 to 5 and 3 to 6.

Def. of function: In this association each element of X has been associated with some element of Y. Moreover no element of X has been associated with more than one element of Y. Such an association between element X and Y is called a function.

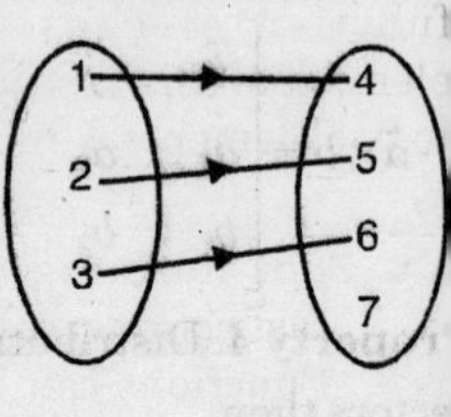

Remark: 1. Any association between the elements of sets A and B need not be a function *e.g* In above figure where X = {1, 2, 3} and Y = {4, 5, 6, 7}

if we associate 1 to 4 and 2 to 5, then this association is not a function because 3 (an element of X) has been left unassociated.

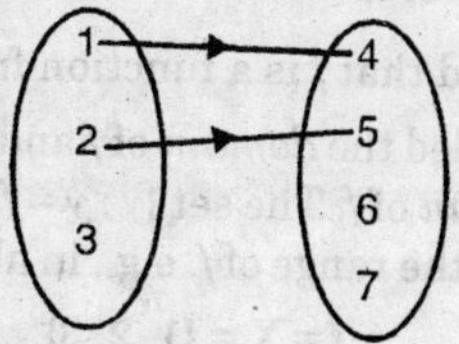

If we associate 1 to 4 and 5; 2 to 6 and 3 to 7. Even this association is not a function because an element of X is associated to two elements 4 and 5 of Y.

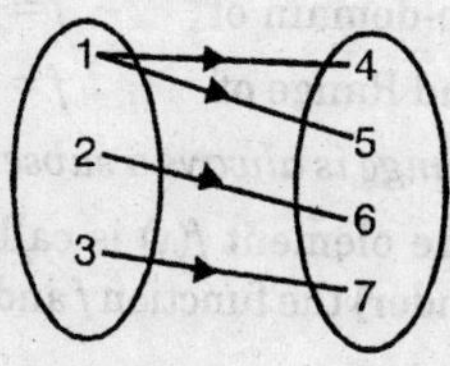

2. For defining a function there are no restrictions on the set Y.
3. Functions can be defined from set X to X also.

Alternate def. of function. A function *f* from a non-empty set X to a set Y is a rule which associates to each element x in X a unique element y in Y.

Notations: The unique element of Y which f associates with x in X is defined by $f(x)$.

$$f : X \longrightarrow Y. \; X \xrightarrow{f} Y$$

usually denoted that f is a function from X to Y. The set X is called the *domain* of f and the set Y is called *co-domain* of f. The set $\{y : y = f(x)$ for some $x \in X)$ is called the range of f, e.g., in above figure.

Domain of $f = X = \{1, 2, 3\}$

Co-domain of $f = Y = \{4, 5, 6, 7\}$

and Range of $f = \{4, 5, 6\} \subset Y$.

Range is always a subset of the co-domain.

The element $f(x)$ is called the image of x by (or under) the function f and x is called the *pre-image*.

Note:

(*i*) The word function is also replaced by mapping or correspondence or transformation.

(*ii*) The function : $f : X \to Y$ is denoted by $y = f(x)$

Here x ranges over X and is called independent variable and y ranges over a subset of Y and is called dependent variable.

(*iii*) If f is a function and x is an object in the domain of f then the image $f(x)$ of x under f is called the value of f at x.

Real Functions: Let $f: A \to B$ be a function from a non empty set A into non-empty set B. In general, the elements of set A and B need not to be numbers. In particular, if A and B are sub set of R (the set of real numbers) then the function f is called a real function.

Value of a function at a point: Let $y = f(x)$ be a function. For any real number 'a' in the domain of f, the value of y corresponding to $x = a$ is called the value of the function at a and is denoted by $f(a)$. For example if $f(x) = x^2 + 7$ then $f(4)$ = value of function at $x = 4$ is equal to $4^2 + 7 = 23$.

Classification of functions

(*i*) **Algebraic functions:** A function which is obtained by a finite number of algebraic operations (e.g., addition, subtraction, multiplication, division, raising to powers, extracting roots etc.) on identity, and constant functions, is called an algebraic function.

Thus, $4x^2 + 7x - 8, \sqrt{x} + \frac{6}{x}$, etc. are all algebraic functions.

Transcendental function: A function which is not algebraic, is called transcendental function. These functions may be

(a) *Trigonometric,* such as $\sin x$, $\tan x$, etc.

(b) *Inverse-Trigonometric,* such as $\sin^{-1}x$, $\cos^{1} x$, etc.

(c) *Logarithmic,* such as $\log_{10}x \log_{a} x$, etc.

(d) *Exponential,* such as 3^x, e^x, $(\cos x)^{\sin x}$, etc.

Explicit and Implicit functions

(a) *Explicit function.* A variable y is said to be an explicit function of another variable x when the value of y is directly expressed in terms of x.

If $y = 4x^2 + 7x + 8$ then y is an explicit function of x.

(b) *Implicit function.* When the variables x and y are expressed in a functional relation, then either variable is an implicit function of the other.

If $3x^2 + 7y^2 + 3axy + 4bx = 0$, then x is an implicit function of y and vice-versa.

Even and odd functions

(a) *Even function.* A function 'f' is said to be even function of x if it remains unaltered in magnitude as well as in sign when x is replaced by $-x$.

$$f \text{ is even if } f(-x) = f(x).$$

(b) *Odd function.* A function 'f' is said to be an odd function of x if it remains unaltered in magnitude but changes in sign when x is replaced by $-x$.

$$f \text{ is odd if } f(-x) = -f(x).$$

(i) **Polynomial function.** A function 'f' defined by

$$f(x) = a_0x^n + a_1x^{n-1} + a_2x^{n-2} + ... + a_{n-1}x + a_n$$

where n is a positive integer and $a_0, a_1...a_{n-1}, a_n$ are constants and $a_0 \neq 0$ is called a polynomial function of nth degree. Thus $5x^3 - 6x^2 + 7x + 1$, $4x^4 - 3x^2 + 1$ are polynomial functions of degree 3 and 4 respectively.

when $n = 0$, it is a constant function

when $n = 1$, it is a linear function,

when $n = 2$, it is a quadratic function,

when $n = 3$, it is a cubic function

when $n = 4$, it is a biquadratic function etc.

Some standard functions and their graphs

(1) **Constant function:** A function of the type $y = f(x) = k$ is called a constant function

where k is a fixed real number. The graph of this function is a straight line parallel to x-axis.

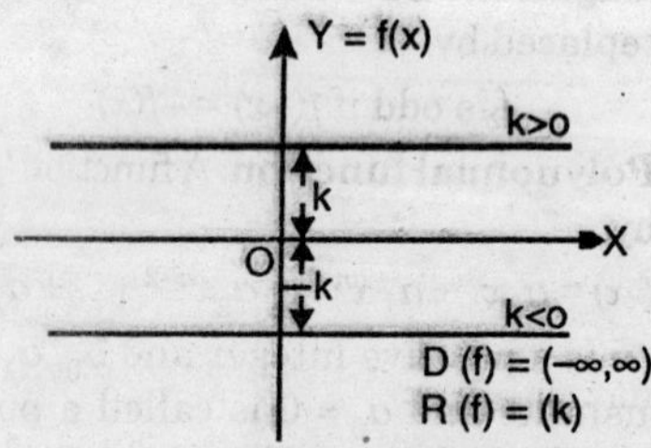

(2) Identity function: The function $y = f(x) = x$ is called the identity function. The graph of $y = f(x) = x$ can also be drawn by finding the values of y for any two different values of x.

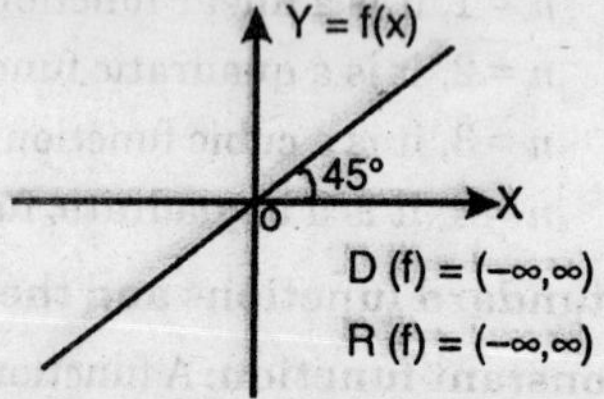

(3) Linear function: A function of the type $y = f(x) = ax + b$, $a \neq 0$ is called a linear function. Thus, graph of $y = f(x) = 2x + 3$ is a straight line.

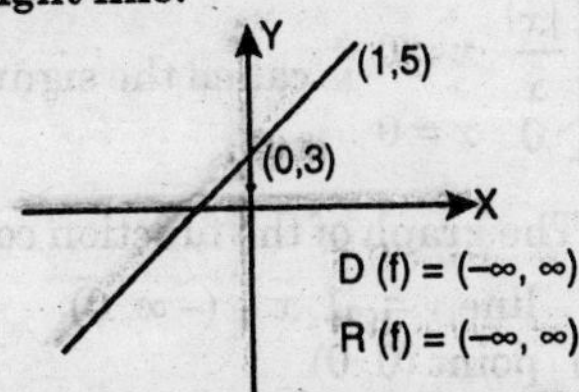

(4) Absolute value function. The function $y = f(x) = |x|$ is called the absolute value function. For every real value of x, the value of $y = |x|$ is unique.

The graph of the function cosists of

(*i*) line $y = -x$, $x \in (-\infty, 0)$

(*ii*) point (0, 0)

(*iii*) line $y = x$, $x \in (0, \infty)$

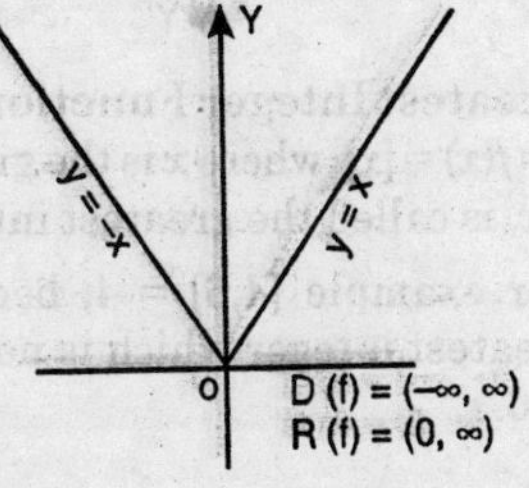

(5) Signum function: The function $f = f(x)$

$$= \begin{cases} \dfrac{|x|}{x} & x \neq 0 \\ 0 & x = 0 \end{cases}$$ is called the signum function.

$\therefore$ The graph of the function consists of

(*i*) line $y = -1, x \in (-\infty, 0)$

(*ii*) point $(0, 0)$

(*iii*) line $y = 1, x \in (0, \infty)$

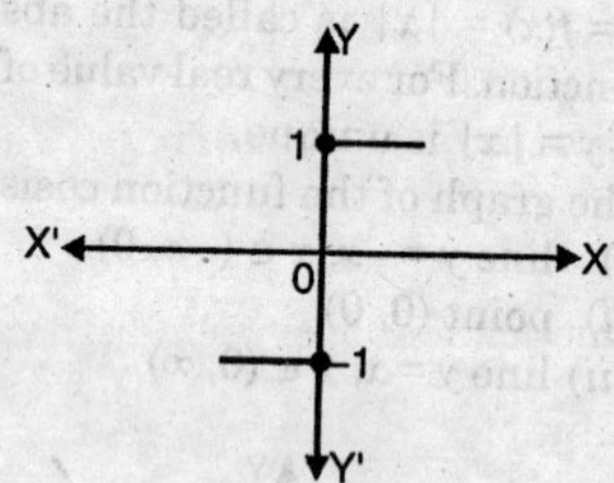

(6) Greatest Integer Function: The function $y = f(x) = [x]$. where x is the greatest integer, $\leq x$, is called the greatest integer function,

For example $[4,6] = 4$, because 4 is the greatest integer which is not greater than 4, 6.

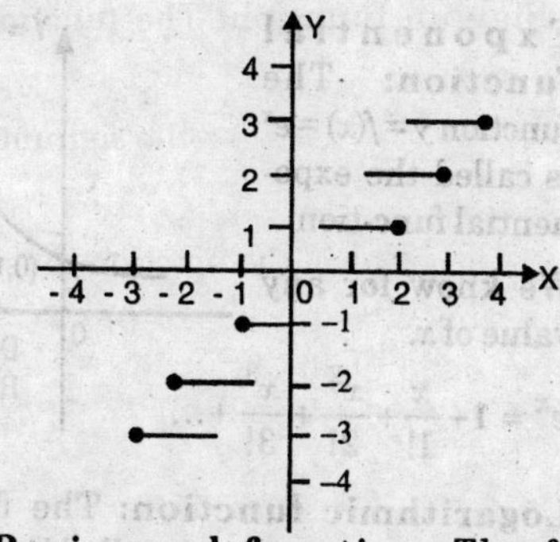

(7) Reciprocal function: The function $y = f(x) = \frac{1}{x}$, $x \neq 0$ is called the reciprocal function. For every non-zero value of x, the value of $y = \frac{1}{x}$ *is unique.*

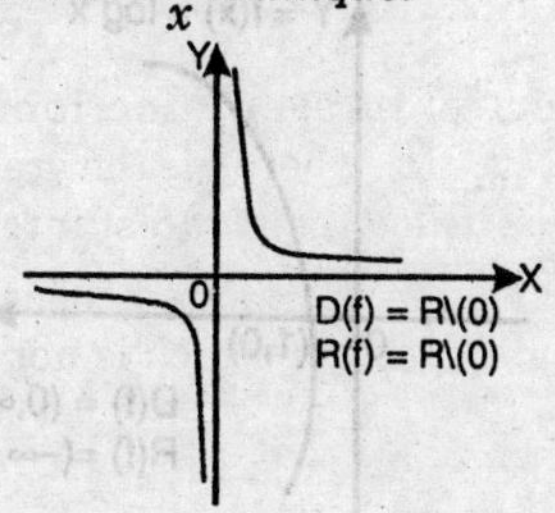

(8) Exponential function: The function $y = f(x) = e^x$ is called the exponential func-tion.

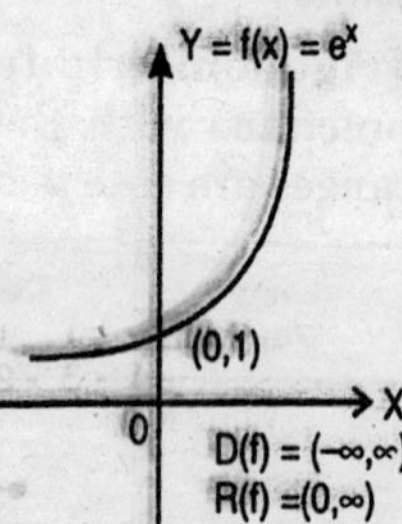

We know for any value of x.

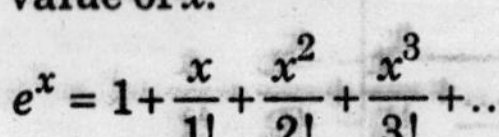

$$e^x = 1 + \frac{x}{1!} + \frac{x^2}{2!} + \frac{x^3}{3!} + \ldots$$

(9) Logarithmic function: The function $y = f(x) = \log x, x \in (0, \infty)$ is called logarithmic function

For $0 < x < \infty$

$y = \log x$ iff $x = e^y$

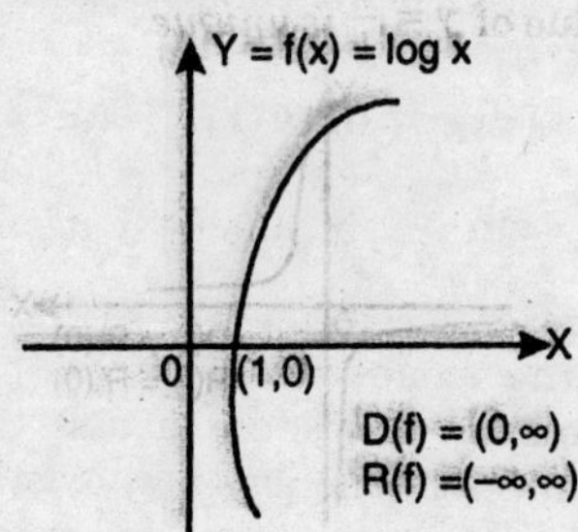

Trigonometric function: The trigonometric functions with their respective domains and ranges are enlisted below :

	Circular Function	Domain	Range
(*i*)	$\sin x$	R	$[-1, 1]$
(*ii*)	$\cos x$	R	$[-1, 1]$
(*iii*)	$\tan x$	$R-(2n+1)\frac{\pi}{2}$	R
(*iv*)	$\cot x$	$R-n\pi$	R
(*v*)	$\sec x$	$R-(2n+1)\frac{\pi}{2}$	$(-\infty, -1] \cup [1, \infty)$
(*vi*)	$\text{cosec } x$	$R-n\pi$	$(-\infty, -1] \cup [1, \infty)$

where R is the set of all real numbers and n is any integer.

Inverse Trigonometric function: As $\sin\frac{\pi}{6} = \frac{1}{2} = \sin\frac{5\pi}{6}$, it follows that $\sin x$ is not one-one. In fact, each of the trigonometric function assumes the same value at infinitely many points. We shall restric the domain suitably so that their inverse functions may exists.

(i) Graph of $y = \sin^{-1} x$

Domain $= -1 \le x \le 1$

Range $= -\frac{\pi}{2} \le y \le \frac{\pi}{2}$

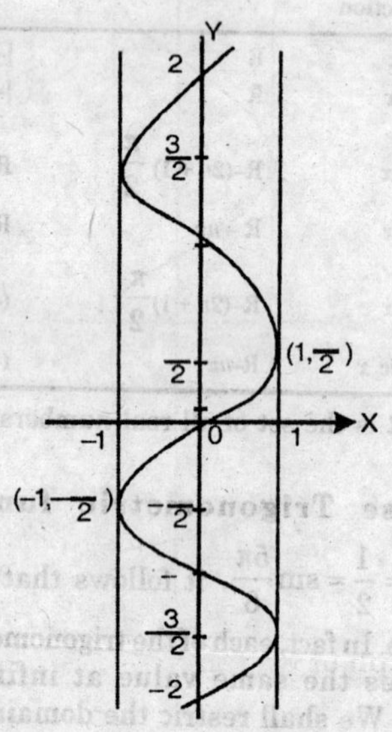

(*ii*) **Graph of $y = \cos^{-1} x$**

Domain $= -1 \le x \le 1$

Range $= 0 \le y \le \pi$

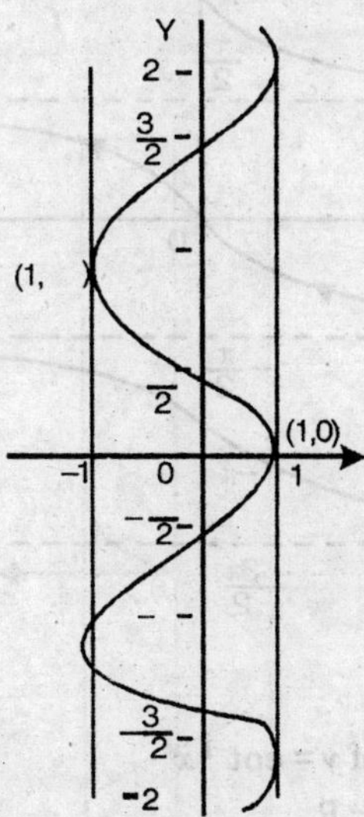

iii) **Graph of $y = \tan^{-1} x$**

Domain = R, Range $= \frac{\pi}{2} < y < \frac{\pi}{2}$

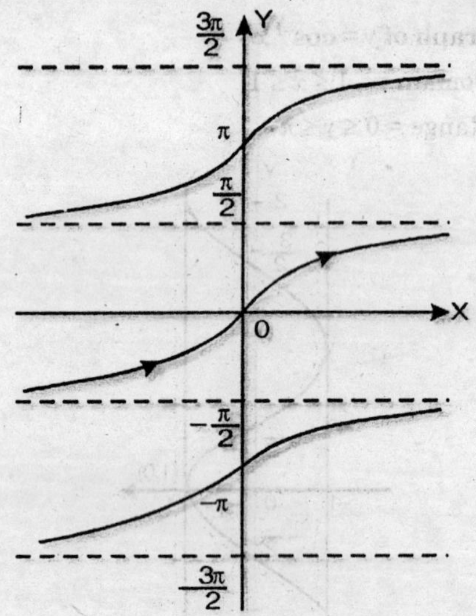

(*iv*) Graph of $y = \cot^{-1} x$

Domain = R

Range = $0 < y < \pi$

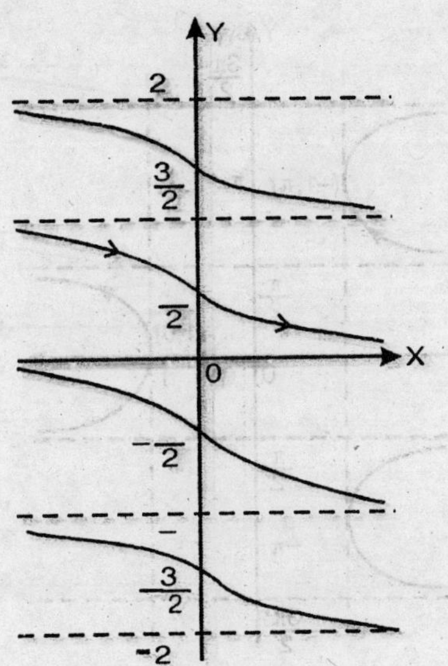

(v) Graph of $y = \sec^{-1}x$

Domain	Range
$x \le -1$	$\frac{\pi}{2} < y \le \pi$
$x \ge 1$	$0 \le y < \frac{\pi}{2}$

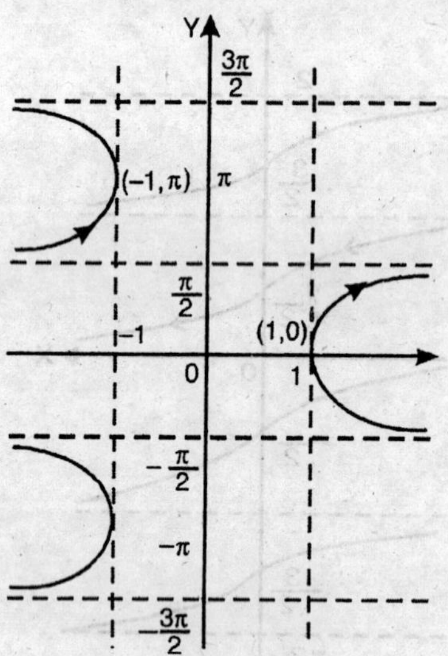

(vi) Graph of $y = \text{cosec}^{-1}x$

Domain	Range
$x \leq -1$	$-\frac{\pi}{2} \leq y < 0$
$x \geq 1$	$0 < y \leq \frac{\pi}{2}$

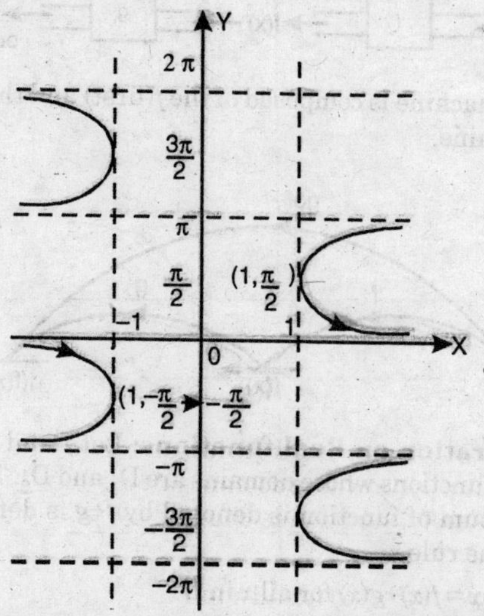

Composite Function: Let $f: A \to A$ and $g: B \to C$ then the composite of the functions f and g denoted by gof or gf is mapping $gof: A \to C$ such that

$(gof)(x) = g\,[f(x)],\ \forall\ x \in A$

$(gof)(x)$ is defined whenever both $f(x)$ and $g(x)$ are defined. We can denote it by a machine or arrow diagram.

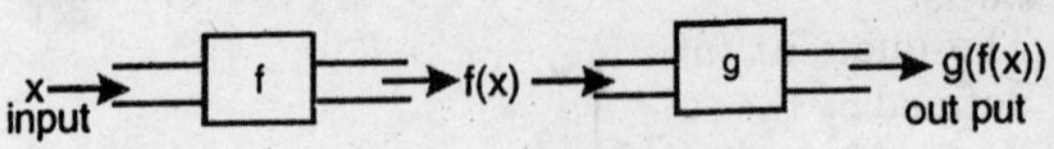

gof machine is composed of the *f*(first) and then *g* machine.

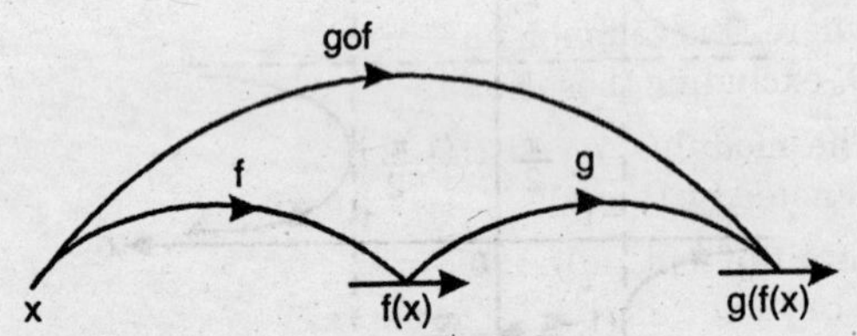

Operation on Real functions: Let *f* and *g* be two functions whose domains are D_1 and D_2. Then the sum of function is denoted by $f+g$ is defined by the rule

$(f+g)x = f(x)+g(x)$ for all x in D

where $D = D_1 \cap D_2$ is the common part of the two domains D_1 and D_2 and $D_1 \cap D_2 \neq \phi$

Difference and product functions denoted by $f-g$ and fg one defined by the rules

$(f-g)x = f(x)-g(x)$ for all x in $D = D_1 \cap D_2$

$(fg)(x) = f(x).\, g(x)$ for all x in $D = D_1 \cap D_2$.

The quotient function f by g denoted by f/g is defined by the rule.

$$\left(\frac{f}{g}\right)(x)=\frac{f(x)}{g(x)} \text{ for all } x \text{ in D.}$$

where D is common part of two domains D_1 and D_2 excluding those points at which $g(x) = 0$.

The modulus $|f|$ of the function f can also be denoted by the rule $|f|(x) = |f(x)|$

Limit of a function: Meaning of x approaches a or "x tends to a". The expression x tends to a, means, the set of infinite (real) values of x which are very close to a and are slightly greater than a and slightly less than a but $x \neq a$.

Limit of a function. Let

$$f(x)=\frac{x^2-1}{x-1} \text{ when } x \neq 1$$

$$=\frac{(x+1)(x-1)}{x-1}=x+1$$

Let x approach 1 through values less than 1,

Putting $\quad x = .9, .99, .999$

$$f(x) = 1.9, 1.99, 1.999.... \quad ...(i)$$

Let x approach 1 through values greater than 1,

Putting $x = 1.1, 1.01, 1.001, \ldots$

$f(x) = 2.1, 2.01, 2.001, \ldots$ ..(*ii*)

Here, we note that as x takes values nearer and nearer to 1, remaining always less than 1 [as in (*i*) or greater than 1 [as in (*ii*), $f(x)$ takes values which nearer and near to 2. The difference between $f(x)$ and 2 i.e..., $|f(x)-2|$ can be made as small as we please by giving to x values close to 1. We say $f(x)$ tends to the limit 2 as x tends to 1.

A function $f(x)$ is said to have a limiting values (approx. value) l as x tends to a if the numerical difference between $f(x)$ and l *i.e.,,* $|f(x)-l|$ can be made as small as we please by giving to x values very close to a but not equal to a.

(*i*) *Left handed and right handed limit.* The limiting value of $f(x)$ as $x \to a^+$ is called the right hand limit of $f(x)$ and is written as

$$\underset{x \to a^+}{\text{Lt}} f(x).$$

The limiting value of $f(x)$ as $x \to a^-$ is called the left hand limit of $f(x)$ and is writen as

$$\underset{x \to a^-}{\text{Lt}} f(x).$$

(ii) *Existence of the limit of a function* $\underset{x \to a^+}{\text{Lt}} f(x)$ is said to exist if the left-handed and right-handed limits both exists and are equal.

Thus if $\underset{x \to a^-}{\text{Lt}} f(x) = \underset{x \to a^+}{\text{Lt}} f(x) = l$, only then we say that $\underset{x \to a}{\text{Lt}} f(x)$ exists and is $= l$

(iii) *Distinction between the value and the limit of a function.* The value of a function $f(x)$ at $x = a$ is obtained by putting $x = a$. The limit of the function $f(x)$ as $x \to a$ is obtained by considering the values of x very close to a.

Thus $\underset{x \to a}{\text{Lt}} f(x)$ may exist even if the function is not defined at $x = a$.

Theorems on Limits

Theorem 1. The limit of a constant quantity is the constant itself.

i.e., $\underset{x \to a}{\text{Lt}}\, c = c$

where c is a constant.

Theorem 2. The limit of the sum of two (or more) functions is equal to the sum of their limits.

$$i.e., \quad \underset{x \to a}{\text{Lt}} [\psi(x) + \varphi(x)] = \underset{x \to a}{\text{Lt}}\, \psi(x) + \underset{x \to a}{\text{Lt}}\, \varphi(x)$$

Theorem 3. The limit of the difference of two functions is equal to the difference of their limits.

i.e., $\underset{x\to a}{\text{Lt}}\,[\psi(x) - \varphi(x)] = \underset{x\to a}{\text{Lt}}\,\psi(x) - \underset{x\to a}{\text{Lt}}\,\varphi(x)$

Theorem 4. The limit of the product of two functions is equal to the product of their limits.

i.e., $\underset{x\to a}{\text{Lt}}\,[\psi(x)\,\varphi(x)] = [\underset{x\to a}{\text{Lt}}\,\psi(x)]\,[\underset{x\to a}{\text{Lt}}\,\varphi(x)]$.

Theorem 5. The limit of the quotient of two functions is equal to the quotient of their limits, provided the limit of the denominator is not zero.

$$\underset{x\to a}{\text{Lt.}}\,\frac{\varphi(x)}{\psi(x)} = \frac{\underset{x\to a}{\text{Lt.}}\,\phi(x)}{\underset{x\to a}{\text{Lt.}}\,\psi(x)}$$

provided $\underset{x\to a}{\text{Lt.}}\,\psi(x) \neq 0$.

Theorem 6. The limit of the product of a constant and a function is equal to the product of the constant and limit of the function.

i.e., $\underset{x\to a}{\text{Lt.}}\,[c.\,\psi(x)] = c\,[\underset{x\to a}{\text{Lt.}}\,\psi(x)]$,

where c is a constant and $\psi(x)$ is any function of x.

Methods for finding the limits of a function

Ist Method

(By factorization)

When $f(x)$ is of the form $\dfrac{g(x)}{h(x)}$

(i) Factorise $g(x)$ and $h(x)$ and cancel the common factors.

(ii) Put the value of x.

Example. *Evaluate* $\underset{x \to a}{\text{Lt.}} \dfrac{\sqrt{x}-\sqrt{a}}{x-a}$

Sol. $$\underset{x \to a}{\text{Lt.}} \frac{\sqrt{x}-\sqrt{a}}{x-a} = \underset{x \to a}{\text{Lt.}} \frac{\sqrt{x}-\sqrt{a}}{\left(\sqrt{x}-\sqrt{a}\right)\left(\sqrt{x}+\sqrt{a}\right)}$$

[Factorising]

$$= \underset{x \to a}{\text{Lt.}} \frac{1}{\sqrt{x}+\sqrt{a}}.$$

$[\because x \to a \therefore x-a \neq 0]$

$$= \frac{1}{\sqrt{a}+\sqrt{a}} = \frac{1}{2\sqrt{a}}$$

2 nd Method (*By substitution*)

To evaluate $\underset{x\to a}{\text{Lt.}}\dfrac{g(x)}{h(x)}$

(*i*) Put $x = a + h$, where $(h \neq 0)$ is very small.
As $x \to a$, then $h \to 0$

(*ii*) Simplify the numerator and denominator so as to cancel h throughout as $h \neq 0$.

(*iii*) Put $h = 0$ and get the required limit.

Example: *Evaluate* $\underset{x\to 1}{\text{Lt.}}\dfrac{x^n - 1}{x - 1}$

Sol. Put $x = 1 + h$, where h is very small.

Since $x \to 1 \therefore 1 + h \to 1$ *i.e.*, $h \to 0$

$$\therefore \underset{x\to 1}{\text{Lt.}}\frac{x^n - 1}{x - 1} = \underset{h\to 0}{\text{Lt.}}\frac{(1+h)^n - 1}{(1+h) - 1}$$

$$= \underset{h\to 0}{\text{Lt.}}\frac{\left[\left(1 + nh + \dfrac{n(n-1)}{2!}h^2 + \ldots\right) - 1\right]}{h}$$

[Binomial theorem)

$$= \underset{h \to 0}{\text{Lt.}} \frac{nh + \frac{n(n-1)}{2!}h^2 + \dots}{h}$$

$$= \underset{h \to 0}{\text{Lt.}} \left[n + \frac{n(n-1)}{2!}h + \dots \right]$$ [Cancelling h as $h \neq 0$.

$= n + 0 + 0 \dots$

$= n.$

3rd Method. In this case we rationalise the factor which contains radical sign. Then simplfy.

Example: *Evaluate* $\underset{x \to 0}{\text{Lt.}} \frac{\sqrt{1+x} - \sqrt{1-x}}{x}$

Sol. $\lim\limits_{x \to 0} \frac{\sqrt{1+x} - \sqrt{1-x}}{x} \times \frac{\sqrt{1+x} + \sqrt{1-x}}{\sqrt{1+x} + \sqrt{1-x}}$

[Rationalising the numerator]

$$= \underset{x \to 0}{\text{Lt}} \frac{(1+x) - (1-x)}{x\left[\sqrt{1+x} + \sqrt{1-x}\right]}$$

$$= \underset{x \to 0}{\text{Lt}} \frac{2x}{x\left[\sqrt{1+x} + \sqrt{1-x}\right]}$$

$$= \underset{x \to 0}{\text{Lt}} \frac{2}{\sqrt{1+x}+\sqrt{1-x}}$$

$$= \frac{2}{\sqrt{1+0}+\sqrt{1-0}} = \frac{2}{1+1} = \frac{2}{2} = 1$$

4th Method. $\underset{x \to \infty}{\text{Lt}} f(x) = \frac{ax^2+bx+c}{dx^2+ex+f}$

(*i*) Divide the numerator and denominator by the highest power of x of occuring in $f(x)$.

(*ii*) Use the idea of $\frac{1}{x}, \frac{1}{x^2} \ldots. \to 0$ *as* $x \to \infty$.

Example: *Evaluate* $\underset{x \to \infty}{\text{Lt}} \frac{4x^2+5x+6}{3x^2+4x+5}$

Sol. $\underset{x \to 0}{\text{Lt}} \frac{4x^2+5x+6}{3x^2+4x+5}$

$$= \underset{x \to \infty}{\text{Lt}} \frac{4+\frac{5}{x}+\frac{6}{x^2}}{3+\frac{4}{x}+\frac{5}{x^2}}$$

[Dividing the numerator and denominator by x^2]

$$= \frac{4+0+0}{3+0+0} \quad \left[\because \frac{5}{x}, \frac{6}{x^2}, \frac{4}{x}, \frac{5}{x^2} \text{ all} \to 0 \text{ at } x \to \infty\right]$$

$$= \frac{4}{3}$$

Some important limits:

(*i*) $\underset{\theta \to 0}{\text{Lt}} \cos\theta = 1$

(*ii*) $\underset{\theta \to 0}{\text{Lt}} \frac{\sin\theta}{\theta} = 1$

(*iii*) $\underset{\theta \to 0}{\text{Lt}} \frac{\tan\theta}{\theta} = 1$

Cor 1. $\underset{\theta \to 0}{\text{Lt}} \sec\theta = 1$

Cor 2. $\underset{\theta \to 0}{\text{Lt}} \sin\theta = 0$

Cor 3. $\underset{\theta \to 0}{\text{Lt}} \frac{\sin^{-1}\theta}{\theta} = 1$

Cor 4. $\underset{\theta \to 0}{\text{Lt}} \frac{\tan^{-1}\theta}{\theta} = 1$

(*iv*) $\underset{n \to 0}{\text{Lt}} \left(1+\frac{1}{n}\right)^n = e$

(v) $\underset{n\to\infty}{\text{Lt}} (1+n)^{1/n} = e$

(vi) $\underset{x\to 0}{\text{Lt}} \dfrac{a^x - 1}{x} = \log_e a$

(vii) $\underset{x\to 0}{\text{Lt}} \log \dfrac{(1+x)}{x} = 1$

(viii) $\underset{x\to 0}{\text{Lt}} \dfrac{x^n - a^n}{x-a} = na^{n-1}$

where n is an integer, $n \neq 0$.

(ix) $\underset{x\to 0}{\text{Lt}} \dfrac{e^x - 1}{x} = 1$

Continuity of Functions

(i) A function $y = f(x)$ is said to be continuous in an interval if its graph is obtained by moving the pen over the various points without lifting it from the paper. On the other hand, if we have to lift the pen in drawing the

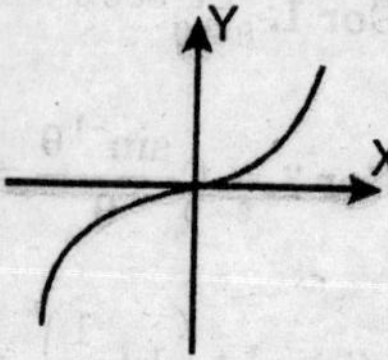

Y

X

graph, then the function is said to be discontinuous. The graph of a discontinuous function is broken. It has gaps or jumps at points of discontinuity.

(*ii*) A function $f(x)$ is said to be continuous at $x = a$, if

(a) $\underset{x \to a}{\text{Lt}}\, f(x)$ exists *i.e.* both

$\underset{x \to a^-}{\text{Lt}}\, f(x)$ and $\underset{x \to a^+}{\text{Lt}}\, f(x)$ exists and are equal

(b) $f(a)$ exists.

(c) $\underset{x \to a}{\text{Lt}}\, f(x) = f(a)$

If a function $f(x)$ is not continuous at $x = a$ it is said to be discontinuous at $x = a$.

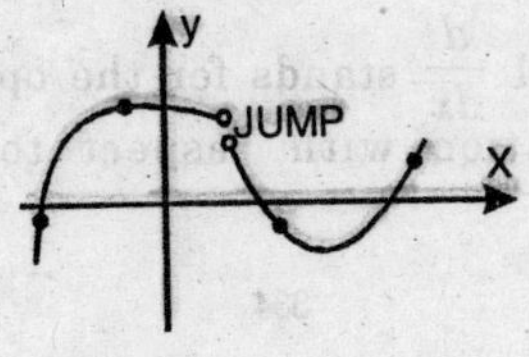

21

DIFFERENTIATION AND THEIR APPLICATION

Derivative or differential coefficient of a function: If $y = f(x)$ is a function of x and δx, a small change in the value of x, then $\underset{\delta x \to 0}{\text{Lt}} \frac{\delta y}{\delta x}$ $= \underset{\delta x \to 0}{\text{Lt}} \frac{f(x+\delta x)-f(x)}{\delta x}$ if it exists, is called the differential coefficient or derivative of $y = f(x)$ and is denoted by $\frac{dy}{dx}$ or $f'(x)$.

Note 1. $\frac{dy}{dx}$ is read as 'dee y by dee x'.

The symbol $\frac{d}{dx}$ stands for the operation of differentiation with respect to x. Thus $\frac{dy}{dx} = \frac{d}{dx}(y)$

Note 2. The process of finding $\frac{dy}{dx}$ is called differentiation

Note 3. The process of finding the derivative of a function by using the definition of derivative as a limit is called *differentiation from first principles* or *differentiation ab-initio* or *differentiation by delta method.*

Work Rule

1. Denote the given function by y i.e., let $y = f(x)$.
2. Let δx be a small change in x and δy then the corresponding change in y, so that $y + \delta y = f(x + \delta x)$.
3. Find δy by subtracting y from $y + \delta y$. Thus $\delta y = f(x + \delta x) - f(x)$.
4. Divide both sides by δx to obtain the difference quotient
$$\frac{\delta y}{\delta x} = \frac{f(x+\delta x) - f(x)}{\delta x}.$$
5. Find the limit of $\frac{\delta y}{\delta x}$ as $\delta x \to 0$
$$\frac{dy}{dx} = \underset{\delta x \to 0}{\text{Lt}} \frac{\delta y}{\delta x}$$

Derivative of x^n where $n \in R$

$$\frac{d}{dx}\left(x^n\right) = nx^{n-1}$$

[Write power before x and subtract one from the power.]

e.g. $\frac{d}{dx}\left(x^8\right) = 8x^{8-1} = 8x^7$

Note. $\frac{d}{dx}(x) = 1$ *i.e.* rate of change of any variable w.r.t itself is 1.

Caution. $\frac{d}{dx}(y)$ is $\frac{dy}{dx}$ and not 1.

Fundamental theorems on differentiation:

Theorem 1. Derivative of a constant is zero.

i.e. $\frac{d}{dx}(c) = 0$

e.g., $\frac{d}{dx}(\pi) = 0; \frac{d}{dx}(9) = 0.$

Theorem 2. The derivative of the product of a constant and a function is equal to the product of the constant and the derivative of the function.

i.e. $$\frac{d}{dx}[c\,f(x)] = c\frac{d}{dx}[f(x)]$$

e.g., $$\frac{d}{dx}\left(5x^7\right) = 5\frac{d}{dx}\left(x^7\right) = 5\left(7x^{7-1}\right)$$
$$= 5\left(7x^6\right) = 35x^6$$

Theorem 3. The derivative of the algebraic sum of any finite number of functions is the algebraic sum of their derivatives.

i.e. $$\frac{d}{dx}[f(x) + g(x) - h(x) + ...]$$

$$= \frac{d}{dx}f(x) + \frac{d}{dx}g(x) - \frac{d}{dx}h(x) + ...$$

e.g., $$\frac{d}{dx}\left(x^3 - 4x + 8\right)$$

$$= \frac{d}{dx}\left(x^3\right) - \frac{d}{dx}(4x) + \frac{d}{dx}(8)$$

$$= 3x^2 - 4 + 0 = 3x^2 - 4.$$

Theorem 4. An additive constant disappears in differentiation.

i.e. $$\frac{d}{dx}[f(x)+c]=\frac{d}{dx}[f(x)].$$

where c is a constant.

e.g., $$\frac{d}{dx}\left[x^{10}+16\right]=\frac{d}{dx}x^{10}=10x^{9}$$

Theorem 5. Derivative of product of two functions; the differential coefficient of the product of two functions is the sum of the products of each function with the derivative of the other.

i.e., $$\frac{d}{dx}(u.v)=u\frac{dv}{dx}+v\frac{du}{dx}$$

where u and v are differential functions of x.

Working Rule: The differential coefficient of the product of two functions = first function × the differential coefficient of the second + second function × the differential coefficient of the first.

Cor. $$\frac{d}{dx}(uvw)=uv\frac{dw}{dx}+uw\frac{dv}{dx}+vw\frac{du}{dx}$$

where u, v, w are functions of x and their derivative exists.

Theorem 6. The differential coefficient of the quotient of two functions is

i.e., $$\frac{d}{dx}\left(\frac{u}{v}\right)=\frac{v\frac{du}{dx}-u\frac{dv}{dx}}{v^2}$$

where u and v are derivable functins of x

In words

$$=\frac{\text{Denom} \times \text{Derivative of Num} - \text{Num} \times \text{Derivative of Denom}}{(\text{Denom})^2}$$

Derivatives of Trigonometric Functions:

$$\frac{d}{dx}\sin x=\cos x \qquad \frac{d}{dx}\sin u=\cos u\times\frac{du}{dx}$$

$$\frac{d}{dx}\cos x=-\sin x \qquad \frac{d}{dx}\cos u=-\sin u\times\frac{du}{dx}$$

$$\frac{d}{dx}\tan x=\sec^2 x \qquad \frac{d}{dx}\tan u=\sec^2 u\times\frac{du}{dx}$$

$$\frac{d}{dx}\cot x=-\operatorname{cosec}^2 x \qquad \frac{d}{dx}\cot u=-\operatorname{cosec}^2 u\times\frac{du}{dx}$$

$$\frac{d}{dx}\sec x=\sec x\tan x \qquad \frac{d}{dx}\sec u=\sec u\tan u\times\frac{du}{dx}$$

$$\frac{d}{dx}\operatorname{cosec} x=-\operatorname{cosec} x\cot x \qquad \frac{d}{dx}\operatorname{cosec} u=-\operatorname{cosec} u\cot u\times\frac{du}{dx}$$

Note:

(*i*) Differential coefficients of those trigonometrical ratios which begins with "co" are negative.

(*ii*) Never forget to multiply by the differential coefficient of u *i.e.*, angle *e.g.*, $\frac{d}{dx}[\sin 6x]$ $= \cos 6x \times 6 = 6\cos 6x$.

Properties of Inverse Trigonometric Functions:

1. *Some angle can be expressed by different inverse trigonometric functions.* We know that

$$\sin 60° = \frac{\sqrt{3}}{2} \quad \Rightarrow \quad 60° = \sin^{-1}\left(\frac{\sqrt{3}}{2}\right)$$

$$\cos 60° = \frac{1}{2} \quad \Rightarrow \quad 60° = \cos^{-1}\left(\frac{1}{2}\right)$$

$$\tan 60° = \sqrt{3} \quad \Rightarrow \quad 60° = \tan^{-1}\left(\sqrt{3}\right)$$

and so on, $60° = \sin^{-1}\left(\frac{\sqrt{3}}{2}\right) = \cos^{-1}\left(\frac{1}{2}\right)$

$= \tan^{-1}\left(\sqrt{3}\right) = ...$

2. **Inverse property.** We know that if $x = \cos\theta$ then, $\theta = \cos^{-1} x$.

   $\therefore \theta = \cos^{-1} (\cos \theta)$ $(\because x = \cos \theta)$

3. **Principle of reciprocity**

   (*i*) $\operatorname{cosec}^{-1}\frac{1}{x} = \sin^{-1} x$

   (*ii*) $\sec^{-1}\frac{1}{x} = \cos^{-1} x$

   (*iii*) $\cot^{-1}\frac{1}{x} = \tan^{-1} x$

4. ***Inverse trigonometric functions are odd functions within the principle value i.e.,***

   (*i*) $\sin^{-1}(-x) = -\sin^{-1} x$

   (*ii*) $\operatorname{cosec}^{-1}(-x) = -\operatorname{cosec}^{-1} x$

   (*iii*) $\tan^{-1}(-x) = -\tan^{-1} x$

5. ***Some fundamental Formulae***

   (*i*) $\sin^{-1} x + \cos^{-1} x = \frac{\pi}{2}$.

(*ii*) $\tan^{-1} x + \cot^{-1} x = \frac{\pi}{2}$

(*iii*) $\text{cosec}^{-1} x + \sec^{-1} x = \frac{\pi}{2}$

(*iv*) $\tan^{-1} x + \tan^{-1} y = \tan^{-1}\left(\frac{x+y}{1-xy}\right)$

(*v*) $\tan^{-1} x - \tan^{-1} y = \tan^{-1}\left(\frac{x-y}{1+xy}\right)$

(*vi*) $2\tan^{-1} x = \sin^{-1} \frac{2x}{1+x^2} = \cos^{-1} \frac{1-x^2}{1+x^2}$

$$= \tan^{-1} \frac{2x}{1-x^2}$$

6. To express one inverse trigonometric function in terms of another one *i.e.*,

(*i*) $\sin^{-1} x = \cos^{-1} \sqrt{1-x^2} = \tan^{-1} \frac{x}{\sqrt{1-x^2}}$

(*ii*) $\cos^{-1} x = \sin^{-1} \sqrt{1-x^2} = \tan^{-1} \frac{\sqrt{1-x^2}}{x}$

(*iii*) $\text{cosec}^{-1} \frac{1}{x} = \sec^{-1} \frac{1}{\sqrt{1-x^2}} = \cot^{-1} \frac{\sqrt{1-x^2}}{x}$

Derivatives of inverse trigonometric functions:

(i) $\frac{d}{dx}\left(\sin^{-1} x\right) = \frac{1}{\sqrt{1-x^2}}$

$\frac{d}{dx}\left(\sin^{-1} u\right) = \frac{1}{\sqrt{1-u^2}} \cdot \frac{d}{dx}(u)$

(ii) $\frac{d}{dx}\left(\cos^{-1} x\right) = \frac{-1}{\sqrt{1-x^2}}$

$\frac{d}{dx}\left(\cos^{-1} u\right) = \frac{-1}{\sqrt{1-u^2}} \cdot \frac{d}{dx}(u)$

(iii) $\frac{d}{dx}\left(\tan^{-1} x\right) = \frac{1}{1+x^2}$

$\frac{d}{dx}\left(\tan^{-1} u\right) = \frac{1}{1+u^2} \cdot \frac{d}{dx}(u)$

(iv) $\frac{d}{dx}\left(\cot^{-1} x\right) = \frac{-1}{1+x^2}$

$\frac{d}{dx}\left(\cot^{-1} u\right) = \frac{-1}{1+u^2} \cdot \frac{d}{dx}(u)$

(v) $\frac{d}{dx}\left(\sec^{-1} x\right)=\frac{1}{x\sqrt{x^2-1}}$

$$\frac{d}{dx}\left(\sec^{-1} u\right)=\frac{1}{u\sqrt{1-u^2}}.\frac{d}{dx}(u)$$

(vi) $\frac{d}{dx}\left(\text{cosec}^{-1}x\right)=\frac{-1}{x\sqrt{x^2-1}}$

$$\frac{d}{dx}\left(\text{cosec}^{-1}u\right)=\frac{-1}{u\sqrt{1-u^2}}.\frac{d}{dx}(u)$$

Some properties of logarithms:

(i) $\log_a mn = \log_a m + \log_a n$

(ii) $\log_a \frac{m}{n} = \log_a m - \log_a n$

(iii) $\log_a m^n = n \log_a m$

(iv) $\log_a m \log_b a = \log_b m$

(v) $\log_a b . \log_b a = 1.$

Where m, n, a and b are positive real numbers.

Logarithmic Differentiation: If u is a function of x, then

$$\frac{d}{dx}\left(u^n\right) = nu^{n-1}\frac{du}{dx}, \frac{d}{dx}\left(a^u\right) = a^u \log a\frac{du}{dx}$$

In this case, the power is a constant and in the second case, the base is constant.

To differentiate u^v where u and v both are functions of x, we first take logarithms of both sides and then differentiate. This process is called logarithmic differentiation. This process is also useful when the function consists of the product or quotient of a number of functions.

Example. Differentiate $(1+x)^{2x}$ $w.r.t.x$

Sol. Let $y = (1+x)^{2x}$ [Form u^v]

Taking log to both sides

$$\log y = \log (1+x)^{2x} = 2x\log(1+x)$$

$$\left[\left(\log m^n = n\log m\right)\right]$$

Differentiating w.r.t.x, we get

$$\frac{1}{y}\frac{dy}{dx} = 2x\frac{d}{dx}\left[\log(1+x)\right] + \log(1+x).\frac{d}{dx}(2x)$$

$$= 2x.\frac{1}{x+1} + \log(1+x)(2)$$

$$\frac{dy}{dx} = y\left[\frac{2x}{x+1} + 2\log(1+x)\right]$$

Putting the value of y

$$\frac{dy}{dx} = 2(1+x)^{2x}\left[\frac{x}{1+x} + \log(1+x)\right]$$

Function of a function : If y a function of u and u in turn is a function of x then y is called a *function of a function, or a composite function.*

Theorem : If y is a function of u and u in turn is a function of x, then y is a function of x and

$$\frac{dy}{dx} = \frac{dy}{du}.\frac{du}{dx} \quad \text{(Chain Rule)}$$

Note 1. *Extension of Chain Rule*

If $y = f(u)$, $u = g(u)$ and $v = h(x)$

then $\frac{dy}{dx} = \frac{dy}{du}.\frac{du}{dv}.\frac{dv}{dx}$

Note 2. If $y = u^n$ and $u = g(x)$ = a function of x.

$$\Rightarrow \frac{dy}{dx} = nu^{n-1}\frac{d}{dx}(u)$$

Aid to memory. $\frac{d}{dx}$ [A function of $x]^n$ = n(function)$^{n-1}$ × diff. coeff. of the function $w.r.t.x$.

Theorem. $\frac{dy}{dx} \times \frac{dx}{dy} = 1$.

Diff. parametric equations. If x and y be expressed in terms of any variable parameter t then

$$\frac{dy}{dx} = \frac{dy/dt}{dx/dt}$$

Second Derivative : If $y = f(x)$ is a differentiable function of x, then its derivative $\frac{dy}{dx}$ is also a function of x. If the function $\frac{dy}{dx}$ of x is also differentiable, then its derivative is denoted by $\frac{d^2y}{dx^2}$ or $f''(x)$ or $\frac{d^2y}{dx^2}$, is called the seconed derivative of $y = f(x)$. $\frac{d^2y}{dx^2}$ is read as dee two y over dee x squared. $\frac{d^2y}{dx^2}$ is also denoted by y'' or y_2.

Illustration : If $y = \sin^{-1}x$, prove that

$$\frac{d^2y}{dx^2} = \frac{x}{\left(1-x^2\right)^{3/2}}$$

Sol. We have $y = \sin^{-1}x$

$$\frac{dy}{dx} = \frac{1}{\sqrt{1-x^2}} = \left(1-x^2\right)^{-1/2}$$

$$\frac{d^2y}{dx^2} = \frac{d}{dx}\left[\left(1-x^2\right)^{-1/2}\right]$$

$$= -\frac{1}{2}\left(1-x^2\right)^{-3/2}\frac{d}{dx}\left(1-x^2\right)$$

$$= -\frac{1}{2}\frac{1}{\left(1-x^2\right)^{3/2}}(-2x) = \frac{x}{\left(1-x^2\right)^{3/2}}$$

The *n*th Derivative : If $y = f(x)$ is a differentiable function of x, then its nth derivative may or may not exists. For $n(>1) \in \mathrm{N}$, the nth derivative of y exists if the $(n-1)$th derivative of y is differentiable. For example for $n = 3$, the 3rd derivative of $y = f(x)$ exists if $\frac{d^2y}{dx^2}$ is differentiable.

If the 3rd derivative of $y = f(x)$ is differentiable, then we can talk of its fourth derivative. The nth derivative of $y = f(x)$ is denoted by

$$\frac{d^n y}{dx^n} \text{ or } f^{(n)}(x), y_n, D^n y \text{ etc.}$$

Application of Differentiation

Motion in a Straight line: Let O be a fixed point on a straight line OX. Let P be the position at time t and Q, the position of the particle at time $t + \delta t$. Let OP = x and OQ = $x + \delta x$. Therefore displacement of the particle in time $(t + \delta t) - t = \delta t$ is given by PQ = $(x + \delta x) - x = \delta x$. The ratio $\frac{\delta x}{\delta t}$ is called the average velocity of the particle between P and Q.

The limit of average velocity $\frac{\delta x}{\delta t}$ as $\delta t \to 0$ is defined as the velocity of the particle P.

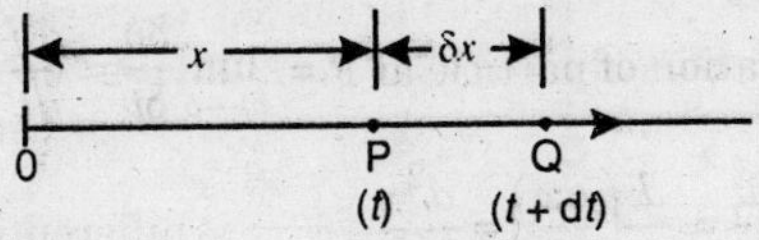

Velocity of particle at P = $\lim_{\delta t \to 0} \frac{\delta x}{\delta t} = \frac{dx}{dt}$

The velocity is denoted by $v = \frac{dx}{dt}$

Note 1. If $v = \frac{dx}{dt} > 0$, then the particle moves in the direction of x increasing, because in the case δx and δt will have same sign.

Note 2. If $v = \frac{dx}{dt} < 0$, then the particle moves in the direction of x decreasing, because δx and δt will have opposite signs.

Note 3. If $v = \frac{dx}{dt} = 0$, then the particle is instantaneously at rest.

Acceleration : The ratio $\frac{\delta v}{\delta t}$ is called the average rate of change of velocity (or average acceleration) of the particle between the points P and Q.

Acceleration of particle at P = $\lim_{\delta t \to o} \frac{\delta v}{\delta t} = \frac{dv}{dt}$

$$\therefore a = \frac{dv}{dt} = \frac{d}{dt}\left(\frac{dx}{dt}\right) = \frac{d^2x}{dt^2}$$

Note 1. If $a = \frac{dv}{dt} > 0$, then the velocity of particle is increasing.

Note 2. If $a = \frac{dv}{dt} < 0$, then the velocity of particle is decreasing.

Motion Under Gravity: The acceleration of the falling body due to gravity towards the centre of earth is denoted by g. Its value is $g = 32$ ft/sec^2 or 9.8 metres/sec^2. For upward motion g is taken as –ve and for downward motion it is taken as +ve.

Rate of change of quantities: Let x and y be any variables and y, a function of x. Therefore, an increased δx in the value of x shall cause an increment δy (say) in the value of y.

We have seen that the ratio $\frac{\delta y}{\delta t}$ is called the average rate of change of y *w.r.t.* x and limit of $\delta y/\delta x$ as $\delta x \to 0$ is the instantaneous rate of change of y *w.r.t.x.*

If x and y are both functions of parameter t, then

$$\frac{dy}{dt} = \frac{dy}{dx} \cdot \frac{dx}{dt}$$

∴ If the rate of change of x w.r.t. t is known, then we can find the value of rate of change of y w.r.t. t

Increasing and Decreasing functions: Let I be an open interval contained in the domain of a real function f.

(*i*) $f(x)$ is called an increasing function on I.

if $x_1 < x_2$ in I $\Rightarrow f(x_1) < f(x_2)$

(*ii*) $f(x)$ is called a decreasing function on I.

if $x_1 < x_2$ in I $\Rightarrow f(x_1) > f(x_2)$

Again

(*iii*) A function f is said to be increasing at a point x_0, if there is an interval $I = (x_0 - h, x_0 + h)$ around x_0 such that for $x_1, x_2 \in I$

$$x_0 < x_2 \Rightarrow f(x_0) < f(x_2)$$

and $$x_1 < x_2 \Rightarrow f(x_1) < f(x_0)$$

(*iv*) A function f is said to be decreasing at a point x_0, if there is an interval I = $(x_0 - h, x_0 + h)$ around x_0 such that x_1, x_2 I

$$x_0 < x_0 \Rightarrow f(x_0) > f(x_2)$$

and $$x_1 < x_0 \Rightarrow f(x_1) > f(x_0)$$

Note 1. The same function can be increasing function in a certain interval and decreasing function in certain other interval.

Note 2. Certain functions are neither increasing nor decreasing in a given interval.

Theorem 1. A differentiable real function $f(x)$ is increasing on an open interval I if and only if $f'(x) > 0$ for all x in I.

Theorem 2. A differentiable real function is decreasing on an interval if $f'(x)$ for all x in I.

Maxima and Minima

1. A function $f(x)$ is said to be maximum at $x = a$ if $f(a)$ is the greatest value of $f(x)$ in the immediate neighbourhood of $x = a$.

   Graphically: $f(x)$ is maximum at A and its maximum value is AM $= f(a)$

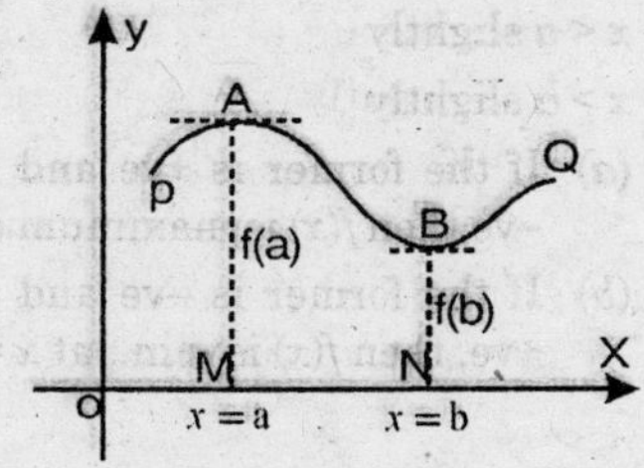

2. A function $f(x)$ is said to be minimum at $x = a$ if (a) is the least value of $f(x)$ in the immediate neighbourhood of $x = a$.

   Graphically : $f(x)$ is minimum at B and its minimum value is BN $= f(b)$.

3. Maximum and minimum values of a function are called extreme values. The points A and B are called stationary points or turning points or points of extreme values.

Necessary and sufficient condition for maximum and minimum values.

Working Rule. For finding the maximum and minimum value of $y = f(x)$.

(*i*) Put $\frac{dy}{dx} = 0$. Solve it for x, giving $x = a, b, c, \ldots$

(*ii*) Select $x = a$, study the sign of $\frac{dy}{dx}$ when

(*i*) $x < a$ slightly

(*ii*) $x > a$ slightly

(*a*) If the former is +ve and later is –ve then $f(x)$ is maximum at $x = a$.

(*b*) If the former is –ve and later is +ve, then $f(x)$ is min., at $x = a$.

(*iii*) Putting these values of x for which $f(x)$ is max. or min. and get the corresponding max. or min. values of $f(x)$.

Use of second derivative Theorem

1. A function $f(x)$ is maximum at $x = a$ if $f'(a) = 0$ and $f''(a) < 0$.
2. A function $f(x)$ is minimum at $x = a$ if $f'(a) = 0$ and $f''(a) > 0$. (+ve)

Working rule to find the max. or min. values

(*i*) Put y = given function $f(x)$ and find $\frac{dy}{dx}$ *i.e.*, $f'(x)$.

(*ii*) Put $\frac{dy}{dx} = 0$ *i.e.*, $f'(x) = 0$ and solve it for x giving $x = a, b, c...$

(*iii*) Select $x = a$, find $\frac{d^2y}{dx^2}$ *i.e.*, $f''(x)$ at $x = a$

(*a*) If $\left(\frac{d^2y}{dx^2}\right)_{x=a}$ i.e., $f''(a)$ is –ve, $x = a$ gives the max. values of the function.

(*b*) If $\left(\frac{d^2y}{dx^2}\right)_{x=a}$ i.e., $f''(a)$ is +ve, $x = a$ gives the min. values of the function.

Rolle's theorem: If a function $f(x)$ is

(*i*) Continuous in the closed interval $[a, b]$ *i.e.* $a \leq x \leq b$

(*ii*) derivable in the open interval (a, b) *i.e.*, $a < x < b$

(*iii*) $f(a) = f(b)$

then there exists at least one point c in the open interval (a, b) (i.e., $a < c < b$) such that $f'(c) = 0$.

Aid to memory

(*i*) Rolle's theorem fails for the function which does not even satisfies one condition.

(*ii*) Every polynomial in x is a continuous function for each x.

$\sin x$, $\cos x$, e^x are continuous for all values of x.

$\log x$ is continuous for all $x > 0$.

(*iii*) If f and g are both continuous on the closed interval $[a, b]$ then $f \pm g$ and fg are also continuous on $[a, b]$

(iv) If $f(x)$ is derivable for every point in a given interval, then it must be continuous in this interval.

i.e., **Derivability** $\Rightarrow$ continuity.

Lagrange's mean value theorem: If a function $f(x)$ is,

(*i*) continuous in the closed interval $[a, b]$ *i.e.*, $a \leq x \leq b$

(*ii*) derivable in the open interval (a, b) *i.e.*, $a < x < b$

then, there exists at least one point c in the open interval (a, b) $[a < c < b]$ such that

$$\frac{f(b) - f(a)}{b - a} = f'(c),$$

Graph of functions: The graph of function $y = f(x)$,

1. Find whether the curve is increasing or decreasing.
   Also, find the turning points, if any
2. Symmetry
   (*i*) Find whether the curve is symmetrical about the x-axis.

   This will happen if no change is affected if y is changed to $-y$. *e.g.*, $y^2 = 4\,ax$ is symmetrical about x-axis.
   [$\because$ only even powers of y occurs]

   (*ii*) Find whether the curve is symmetrical about the y-axis.

This will happen if no change is affected if x is changed to $-x$ *e.g.*, $x^2 = 4\,ay$ is symmetrical about y-axis.

[$\because$ only even powers of x occurs]

Note : $x^2 + y^2 = a^2$ is symmetrical about both axis.

(*iii*) Find whether the curve is symmetrical in opposite quadrants. This will happen if no change is affected if x is changed to $-x$ and y to $-y$ *e.g.*, $xy = k$ is symmetrical in opposite quadrants.

(*iv*) Table. Form a table by taking suitable values of x and y.

(*v*) Plot the above points and join them by free hand drawing so as to get the required rough sketch.

Tangents and Normals: Equations of the tangent and the normal to the curve.

Here, the equation of the curve is

$$y = f(x)$$

$$\frac{dy}{dx} = \text{slope of tangent at } (x, y)$$

$$\therefore \text{ Slope of tangent at } P(x_1, y_1) = \left(\frac{dy}{dx}\right)_{\text{at}(x_1,\, x_2)} = m$$

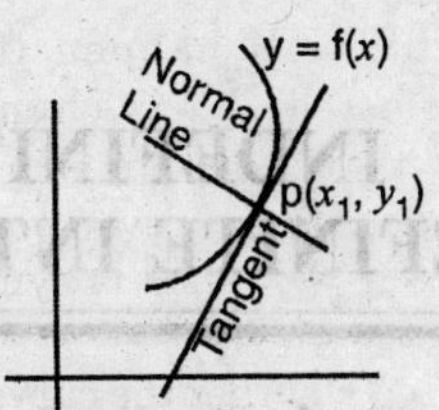

The equation of tangent at (x_1, y_1) is

$y - y_1 = m(x - x_1)$

Slope of normal line at a given point is the negative of the reciprocal of the slope of the tangent line at that point.

Slope of normal at $P(x_1, y_1) = -\frac{1}{m}(m \neq 0)$.

The equation of the normal at (x_1, y_1) is

$$(y - y_1) = -\frac{1}{m}(x - x_1).$$

INDEFINITE AND DEFINITE INTEGRAL

Integration is the inverse operation of differentiation: In differential calculus, we are given a function and we are required to differentiate it. In integral calculus we are required to find a function whose differential coefficient (or derivative) is given.

If the differential coefficient of a function $f(x)$ w.r.t. x is $F(x)$ then we say that the integral (or primitive) of $F(x)$ w.r.t. x is $f(x)$.

Symbolically. If $\frac{d}{dx}[f(x)] = F(x)$ then

$\int F(x)\,dx = f(x)$ and is read as "integral of $F(x)$ w.r.t. x is $f(x)$".

The symbol $\int$ (elongated *S*) is called the sign of integration, $f(x)$ is called the integrand and the process of finding $f(x)$ is called integration.

Constant of integration

If $\frac{d}{dx}[f(x)] = F(x)$ then, $\int F(x)\,dx = f(x) + c$ where c is an arbitrary constant.

The constant c is called the constant of integration. The constant of integration is generally omitted.

Note : Different methods of integrating a function may give different answers apparently, but any two such answers differ only by a constant.

Standard Forms

(*i*) $\int x^n dx = \frac{x^{n+1}}{n+1} [n \neq -1]$

i.e., increase the index of x by 1 and divide by the new index.

(*ii*) $\int \frac{1}{x} dx = \log x$

Note. In (*i*) $n = -1$

$$\int x^{-1}dx = \int \frac{1}{x}dx = \log x$$

(iii) $\int e^{ax}dx = \frac{e^{ax}}{a}; \int e^{x}dx = e^{x}$

(iv) $\int a^{mx}dx = \frac{a^{mx}}{m}; \int a^{x}dx = \frac{a^{x}}{\log a}$

(v) $\int \sin x \, dx = -\cos x$

(vi) $\int \cos x \, dx = \sin x$

(vii) $\int \sec^2 x \, dx = \tan x$

(viii) $\int \text{cosec}^2 x \, dx = -\cot x$

(ix) $\int \sec x \tan x \, dx = \sec x$

(x) $\int \text{cosec}\, x . \cot x = -\text{cosec}\, x$

(xi) $\int 1 \, dx = \int x^0 dx = \frac{x^{0+1}}{0+1} = x$

Important extension of elementary forms:

(*i*) All the results of the above list hold good when x is replaced by $x + a$ [a being a constant] in any formula.

e.g., $$\int (x+a)^n dx = \frac{(x+a)^{n+1}}{n+1} \quad [n \neq -1]$$

$$\int \sec(x+6)\tan(x+6)dx = \sec(x+6)$$

(ii) If x be replaced by $ax + b$ [a and b being constants] on both sides of any standard result of the above table of integrals, the standard form remains true, provided the result on R.H,S. is divided by 'a' the coefficient of x

i.e. $$\int (ax+b)^n dx = \frac{(ax+b)^{n+1}}{(n+1)\,a} \quad [n \neq -1]$$

$$= \int \operatorname{cosec}^2(8x+3)dx = -\frac{\cot(8x+3)}{8}$$

Theorems on integration

Theorem 1. The processes of differentiation and integration neutralises each other *i.e.* $$\frac{d}{dx}\left[\int f(x)dx = f(x)\right]$$

Theorem 2. The integral of the product of a constant and a function is equal to the product of the constant and the integral of the function.

i.e. $$\int c\,f(x)\,dx = c\int f(x)dx$$

e.g. $$\int 5x^3dx = 5\int x^3dx = \frac{5x^{3+1}}{3+1} = \frac{5}{4}x^4$$

Theorem 3. If u, v, w ... (finite number) are functions of x, then

$$f(u+v+w+\ldots)dx = \int udx + \int vdx + \int wdx$$

e.g.,

$$\int\left(a+\frac{b}{x}+\frac{c}{x^2}\right)dx = a\int 1.dx + b\int\frac{1}{x}dx + c\int\frac{1}{x^2}dx$$

$$= ax + b\log x + c\frac{x^{-2+1}}{-2+1} = ax + b\log x - \frac{c}{x}$$

Note: If the degree of the numerator of the integrand is equal to or greater than that of denominator, divide the numerator by the denominator until the degree of the remainder is less than that of denominator *e.g.*

$$\int \frac{x^3}{x+1}\,dx = \int\left(x^2 - x + 1 - \frac{1}{x+1}\right)dx$$

$$= \frac{x^3}{3} - \frac{x^2}{2} + x - \log(x+1)$$

Difinite integral. If $\phi(x)$ be an integral of $f(x)$ then the quantity $\phi(b) - \phi(a)$ is called the definite integral of $f(x)$ between the limits a and b is written as $\int_a^b f(x)dx$. *It is read as integral from a to b of f(x) dx.* 'a' is called the lower limit and b the upper limit, $\phi(b) - \phi(a)$ is written as $[\phi(x)]_a^b$

Thus $\int_a^b f(x)dx = [\phi(x)]_a^b = \phi(b) - \phi(a)$.

Rule to evaluate $\int_a^b f(x)\,dx$

1. Integrate $\int_a^b f(x)\,dx$...

2. In the result first put x = upper limit (b), and then x = lower limit (a).
3. Subtract the second result from the first.

e.g. $$\int_3^4 \frac{dx}{x} = [\log x]_3^4 = \log 4 - \log 3$$

$$\therefore \quad = \log \frac{4}{3} \quad \left[\because \log m - \log n = \log \frac{m}{n}\right]$$

Integration by substitution. Integration of many functions become simple by substitution of a new variable. In other words, many integrals can be evaluated in a simple way by changing the variable of the given integrand, say from the given variable x to the new variable z, the two variable x, z being connected by some relation. This process of integration is called *integration by substitution.*

For example, Let $I = \int \sin^3 x \cos x \, dx$

Put $\sin x = t$, then $\cos x = \frac{dt}{dx}$

[Differentiation w.r.t. x]

or, $\cos x \, dx = dt$

$$\therefore \text{I} = \int t^3 . dt = \frac{t^4}{4} + c$$

$$= \frac{1}{4} \sin^4 x + c$$

Two important forms of integrals

Theorem 1. $\int \frac{f'(x)\, dx}{f(x)} = \log [f(x)]$

Aid to memory. The integral of a fraction whose numerator is differential co-efficient of the denominator is log [denom.]

e.g., $\int \frac{2x+3}{x^2+3x+7} dx$

$$\because \frac{d}{dx}\left(x^2+3x+7\right) = 2x+3$$

$$\therefore \int \frac{2x+3}{x^2+3x+7} dx = \log\left(x^2+3x+7\right)$$

$$\left[\because \int \frac{f'(x)}{f(x)} dx = \log f(x)\right]$$

Theorem 2. $\int [f(x)]^n f'(x)\,dx = \frac{[f(x)]^{n+1}}{n+1}$

Provided $n \neq -1$

Thus if the integrand is the product of a power of a function $f(x)$ and its derivative $f'(x)$ then the integral is obtained by increasing the index of $f(x)$ by 1 and dividing the result by the new index.

e.g., $\int \left(ax^2 + bx + c\right)^5 (2ax + b)\,dx$

[Form $\int [f(x)^n f'(x) dx]$

$$= \frac{\left(ax^2 + bx + c\right)^{5+1}}{5+1}$$

$$= \left[\begin{array}{l} \text{Here } f(x) = \left(ax^2 + bx + c\right) \\ f'(x) = 2ax + b \end{array}\right]$$

Remember. $\int \tan x\,dx = -\log \cos x = \log \sec x.$

Remember. $\int \cot x\,dx = \log \sin x.$

Remember. $\int \text{cosec}\, x\, dx = \log \tan \frac{x}{2}$.

Remember. $\int \text{cosec}\, x\, dx = \log (\sec x - \cot x)$

Remember. $\int \sec x\, dx = \log \tan \left(\frac{\pi}{4} + \frac{x}{2}\right)$.

Remember. $\int \sec x\, dx = \log (\sec x + \tan x)$.

Five Standard Forms

(i) $\int \frac{dx}{\sqrt{a^2 - x^2}} = \sin^{-1} \frac{x}{a}$.

(ii) $\int \frac{dx}{a^2 + x^2} = \frac{1}{a} \tan^{-1} \frac{x}{a}$.

(iii) $\int \frac{dx}{x\sqrt{x^2 - a^2}} = \frac{1}{a} \sec^{-1} \frac{x}{a}$.

(iv) $\int \frac{dx}{\sqrt{a^2 + x^2}} = \sinh^{-1} \frac{x}{a}$

(v) $\int \frac{dx}{\sqrt{x^2 + a^2}} = \cosh^{-1} \frac{x}{a}$

Remember. To integrate a fraction whose numerator is 1 and denominator is a homogeneous function of the second degree in $\cos x$ and $\sin x$.

1. Divide the numerator and denominator by $\cos^2 x$.
2. Put $\tan x = z$.

e.g., $\int \frac{d\theta}{a^2 \sin^2 \theta + b^2 \cos^2 \theta}$ Dividing num. and denom. by $\cos^2\theta$

$$\int \frac{\sec^2 \theta \, d\theta}{a^2 \tan^2 \theta + b^2} = \frac{1}{a^2} \int \frac{dz}{z^2 + \frac{b^2}{a^2}}$$

[**Note.** To make the coeff. of z^2 unity]

$$= \frac{1}{a^2} \int \frac{dz}{z^2 + \left(\frac{b}{a}\right)^2} \left[\text{From} \int \frac{dx}{x^2 + a^2} \text{ Here } a = \frac{b}{a}\right]$$

$$= \frac{1}{a^2} \cdot \frac{1}{\frac{b}{a}} \tan^{-1}\left(\frac{z}{\frac{b}{a}}\right) = \frac{1}{ab} \tan^{-1}\left(\frac{a}{b} z\right)$$

$$= \frac{1}{ab} \tan^{-1}\left(\frac{a}{b}\tan\theta\right) [\because z = \tan\theta]$$

Integration by parts

If u and v be two function of x, then

$$\int uv\, dx = u\int v\, dx - \int \frac{du}{dx}\left(\int v\, dx\right)dx$$

In words. Integral of the product of two functions = Ist function × integral of 2nd-integral of [Diff. coeff. of 1st × Integral of 2nd].

Rule to choose the factor of differentiation or the first function. If one factor of the product is a power of x take it as the first function provided the integral of the second function is handy. If however the integral of the second function is not readily available [in case of inverse circular function or inverse hyperbolic function or a logarithmic function] in that case, take that factor as the first function.

If the integrand is a single inverse circular function (or hyperbolic function) or a single logarithm, take that function as the function and unity (1) as the second function.

$$\int e^x\left[f(x)+f'(x)\right]dx = e^x f(x)$$

$$\int e^x(\sin x+\cos x)\,dx\,\left[\text{Here } f(x)=\sin x, f'(x)=\cos x\right]$$

$$= e^x \sin x$$

$$\int \sqrt{a^2-x^2}\,dx = \frac{x\sqrt{a^2-x^2}}{2}+\frac{a^2}{2}\sin^{-1}\frac{x}{a}$$

$$\int \sqrt{a^2+x^2}\;dx = \frac{x\sqrt{x^2+a^2}}{2}+\frac{a^2}{2}\sinh^{-1}\frac{x}{a}$$

$$\int \sqrt{x^2-a^2}\;dx = \frac{x\sqrt{x^2-a^2}}{2}-\frac{a^2}{2}\cosh^{-1}\frac{x}{a}$$

i.e., for

$$\int \sqrt{a^2-x^2}\;dx, \int \sqrt{a^2+x^2}\;dx, \int \sqrt{x^2-a^2}\;dx$$

$$\text{Integral} = \frac{x\times\text{Integrand}}{2} \cdot \frac{+a^2 \text{or} -a^2 \text{as integrand}}{2}$$

$\times$ integral of [reciprocal of integrand]

Two standard integrals

$$\int e^{ax}\sin bx\,dx=\frac{e^{ax}}{a^2+b^2}(a\sin bx-b\cos bx)$$

$$\int e^{ax}\cos bx\,dx=\frac{e^{ax}}{a^2+b^2}(a\cos bx+b\sin bx)$$

First form

$$\int e^{ax}\sin bx\,dx=\frac{e^{ax}}{\sqrt{a^2+b^2}}\sin\left(bx-\tan^{-1}\frac{b}{a}\right)$$

$$\int e^{ax}\cos bx\,dx=\frac{e^{ax}}{\sqrt{a^2+b^2}}\cos\left(bx-\tan^{-1}\frac{b}{a}\right)$$

2nd form

Important Note. $e^{\log f(x)}=f(x)$

INTEGRATION OF RATIONAL FUNCTIONS

Two standard forms

(i) $\int\frac{dx}{x^2-a^2}\left[x^2>a^2\right]=\frac{1}{2a}\log\frac{x-a}{x+a}$

(ii) $\frac{dx}{a^2-x^2}\left[x^2<a^2\right]=\frac{1}{2a}\log\frac{a+x}{a-x}$

$\int\frac{dx}{ax^2+bx+c}$, a be + ve then.

Case I. When $b^2 > 4ac$ then

$$= \frac{1}{\sqrt{b^2 - 4ac}} \log \frac{2ax + b - \sqrt{b^2 - 4ac}}{2ax + b + \sqrt{b^2 - 4ac}}$$

Case II. When $b^2 < 4ac$ then

$$= \frac{2}{\sqrt{4ac - b^2}} \tan^{-1}\left[\frac{2ax + b}{\sqrt{4ac - b^2}}\right]$$

To integrate $\int \frac{dx}{\text{Quadratic}}$

1. Make the coefficient of x^2 unity by taking the numerical coefficient of x^2 outside.

2. Complete the square in terms containing x by adding and subtracting the square of half the coefficient of x.

3. Use the proper standard form.

Note: If in a numerical problem, the discriminant of quadratic in denominator $[b^2 - 4ac]$ is +ve and a perfect square, factorise the denominator and resolve it into partial fractions then integrate, *e.g.,*

$$\int \frac{dx}{2x^2-2x+1} = \frac{1}{2}\int \frac{dx}{x^2-x+\frac{1}{2}} = \frac{1}{2}\int \frac{dx}{x^2-x+\frac{1}{4}-\frac{1}{4}+\frac{1}{2}}$$

$$= \frac{1}{2}\int \frac{dx}{\left(x-\frac{1}{2}\right)^2+\frac{1}{4}} = \frac{1}{2}\int \frac{dx}{\left(x-\frac{1}{2}\right)^2+\left(\frac{1}{2}\right)^2}$$

$$= \frac{1}{2}\frac{1}{\frac{1}{2}}\tan^{-1}\frac{x-\frac{1}{2}}{\frac{1}{2}} = \tan^{-1}(2x-1).$$

Method to integrate $\int \frac{\text{Linear}}{\text{Quadratic}}\,dx.$

1. Put linear $= \lambda \frac{d}{dx}(\text{quadratic}) + \mu$
2. Equate the coefficients of x and constant terms on both sides to find λ and μ.

The above two steps are taken to break the given fraction into two fractions such that in one the numerator is the derivative of denominator and in the other, numerator is a constant.

e.g., $\int \frac{2x}{x^2+2x+2}\,dx$

Let $2x = \lambda \frac{d}{dx}\left(x^2 + 2x + 2\right) + \mu$

i.e. $2x = \lambda\,(2x + 2) + \mu$

Equating coeffi. x, $2 = 2\lambda \quad \therefore \quad \lambda = 1$

Equating constant term, $0 = 2\lambda + \mu$

$\mu = -2\lambda = -2$

$$\therefore \int \frac{2x}{x^2 + 2x + 2}\,dx = \int \frac{\lambda(2x+2) + \mu}{x^2 + 2x + 2}\,dx$$

$$= \lambda \int \frac{2x+2}{x^2 + 2x + 2}\,dx + \mu \int \frac{dx}{x^2 + 2x + 2}$$

$$= \lambda \log\left(x^2 + 2x + 2\right) + \mu \int \frac{dx}{(x+1)^2 + 1}$$

$$= \log\left(x^2 + 2x + 2\right) - 2\tan^{-1}(x+1)$$

$[\because \lambda = 1, \mu = -2]$

Integration of irrational functions

Rule to integrate

$$\int \frac{dx}{\sqrt{\text{Quadratic}}} \text{ or } \int \sqrt{\text{Quadratic}}\,dx$$

1. Make the coeffi. of x^2 unity by taking its numerical coeffi. outside the square root sign.
2. Complete the square in terms containing x by adding and subtracting the square of half the coeffi. of x.
3. Use proper standard form.

Working rule to calculate the area under a plane curve.

Step I. Make a rough sketch of the graph of the given function.

Step II. Make the region whose area is to be calculated.

Step III. Set up the definite integral for the area in such a way so that limits of integration are so chosen that the independent variable varies throughout the region.

Step IV. Evaluate the definite integral set up in step III.

Reduction formula: A formula which connects an integral which cannot otherwise be evaluated, with and another integral of the same type but of lower degree is called a reduction formula.

In this method, we go on reducing the power till we get a power whose intgral is knwon or which can be integrated easily. Reduction formula is generally obtained by the method of Integration by parts.

(i) Reduction formula for $\int n^x e^{ax}\, dx$ **is**

$$I_n = \frac{1}{a} x^n e^{ax} - \frac{n}{a} I_{n-1}$$

and $I_{n-1} = \frac{1}{a} x^{n-1} e^{ax} - \frac{n-1}{a} I_{n-2}$

[Replacing n by $n-1$]

In this way we get I_{n-2}, I_{n-3}

(ii) Reduction formula for $\int \sin^n x\, dx$ **is**

$$I_n = \frac{-\sin^{n-1} x . \cos x}{n} + \frac{n-1}{n} I_{n-2}$$

(iii) Reduction formula for $\int \cos^n x\, dx$ **is**

$$I_n = \frac{\cos^{n-1} x . \sin x}{n} + \frac{n-1}{n} I_{n-2}$$

(iv) Reduction formula for $\int \tan^n x\, dx$ **is**

$$I_n = \frac{\cos^{n-1} x}{n-1} - I_{n-2}$$

(v) **Reduction formula for $\int \cot^n x\, dx$ is**

$$I_n = \frac{-\cot^{n-1} x}{n-1} - I_{n-2}$$

(vi) **Reduction formula for $\int \sec^n x\, dx$ is**

$$I_n = \frac{\sec^{n-2} x \tan x}{n-1} + \frac{n-2}{n-1} I_{n-2}$$

(vii) **Reduction formula for $\int \operatorname{cosec}^n x\, dx$ is**

$$I_n = \frac{-\operatorname{cosec}^{n-2} x \cot x}{n-1} + \frac{n-2}{n-1} I_{n-2}$$

(viii) **Reduction formula for $\int x^n \sin x\, dx$ is**

$$I_n = -x^n \cos x + n x^{n-1} \sin x - n(n-1) I_{n-2}$$

(ix) **Reduction formula for $\int x^n \cos x\, dx$ is**

$$I_n = x^n \sin x + n\, x^{n-1} \cos x - n(n-1) I_{n-2}$$

(x) Reduction formula for

$\int \sin^m x \cos^n x \, dx$ **is**

$$I_{m,n} = \frac{\cos^{n-1} x . \sin^{m+1} x}{m+n} + \frac{n-1}{m+1} I_{m,n-2}$$

Definite Integral

First Fundamental Theorem of Integral Calculus: Let $f(x)$ be a continuous function of x for $a \le x \le b$ and

$A(x) = \int_o^x f(x)\, dx$, then $A'(x) = f(x)$ for all x in $[a, b]$ and $A(a) = 0$

Second Fundamental Theorem of Integral Calculus: Let f be a continuous function defined on the interval $a \le x \le b$ and ϕ be an anti-derivative of f, then

$$\int_a^b f(x)\, dx = \phi(b) - \phi(a)$$

In words the above theorem tells that

$\int_a^b f(x)\, dx$ = (value of an antiderivative at b, the upper limit) – (value of the same anti-derivative at a, the lower limit)

Note: We often write $\phi(b)-\phi(a)$ as $\left[\phi(x)\right]_a^b$

Definite Integral as a limit of a Sum: If $f(x)$ be a single valued continuous function defined in the interval (a, b) where $a < b$ and the interval (a, b) is divided into n equal parts of each length h so that $nh = b-a$ then we define

$$\int_a^b f(x)\,dx = \lim_{h\to 0} h\left[f(a)+f(a+h)+f(a+2h)+\ldots + f\left(a+\overline{n-1}h\right)\right]$$

OR

$$\int_a^b f(x)\,dx = (b-a)\lim_{h\to\infty}\frac{1}{n}\left[f(a)+f(a+h)+f(a+2h)+\ldots + f\left(a+\overline{n-1}\,h\right)\right]$$

where $h = \dfrac{b-a}{n}$

is called the definite integral of $f(x)$ between the limits $x = a$ and $x = b$.

Some Properties of Definite Integrals

(*i*) $\int_a^b f(x)\,dx = \int_a^b f(t)\,dt$

(ii) $\int_a^b f(x)\,dx = -\int_b^a f(x)\,dx$

(iii) $\int_a^b f(x)\,dx = \int_a^c f(x)\,dx + \int_c^b f(x)\,dx$ if $a < c < b$

(iv) $\int_0^a f(x)\,dx = \int_o^a f(a-x)\,dx$

(v) $\int_{-a}^a f(x)\,dx = 0$, when $f(x)$ is an odd function

$= 2\int_0^a f(x)\,dx$, when $f(x)$ is an even function

(vi) $\int_0^{2a} f(x)\,dx = 2\int_0^a f(x)\,dx$ if $f(2a-x) = f(x) = 0$

if $f(2a-x) = -f(x)$

Definite Integral as Area under the curve: Let $f(x)$ be finite and continuous in $a \le x \le b$. Then area of the region bounded by x–axis, $y = f(x)$ and the ordinates at $x = a$ and $x = b$ (i.e., Area ABCD) is equal to

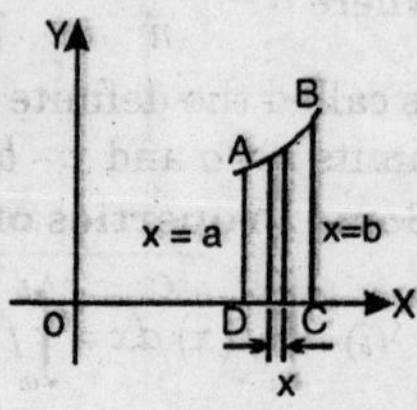

$$\int_a^b f(x)\,dx$$

Remark : In the above fig. we assumed that $f(x) \geq 0$ for all x in $a \leq x \leq b$. However if

(i) $f(x) \leq 0$ for all x in $a \leq x \leq b$ then area bounded by x–axis, the curve $y = f(x)$ and the ordinate $x = a$ to $x = b$ is given by

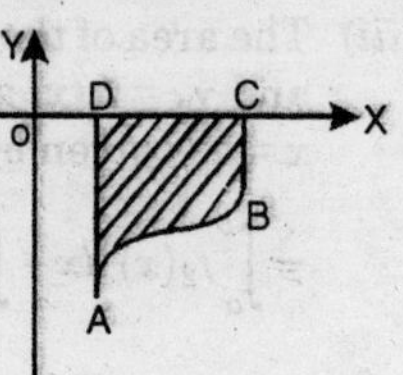

$$= -\int_a^b f(x)\,dx$$

(ii) If $f(x) \geq 0$ for $a \leq x \leq c$ and $f(x) \leq 0$ for $b \leq x \leq c$, then area bounded by $y = f(x)$, x–axis and the ordinates $x = a$, $x = b$ is

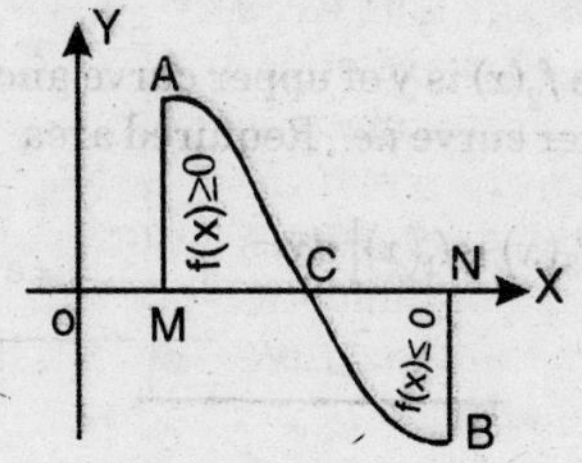

$$= \int_a^c f(x)\,dx + \int_c^b -f(x)\,dx$$

$$= \int_a^c f(x)\,dx - \int_c^b f(x)\,dx$$

(iii) The area of the region bounded by $y_1 = f_1(x)$ and $y_2 = f_2(x)$ and the ordinates $x = a$ and $x = b$ is given by

$$= \int_a^b f_2(x)\,dx - \int_a^b f_1(x)\,dx$$

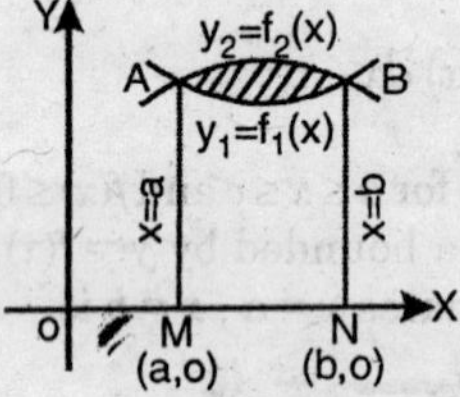

where $f_2(x)$ is y of upper curve and $f_1(x)$ is y of lower curve *i.e.*, Required area

$$= \int_a^b [f_2(x) - f_1(x)]\,dx$$

23 DIFFERENTIAL EQUATIONS

A differential equation is an equation involving independent variable, dependent variable and the derivatives of the dependent variable. In otherwords, if x is an independent variable, and y is dependent on x, then any equation involving x, y and $\frac{dy}{dx}$ is called a differential equation e.g.,

(i) $\frac{dy}{dx} + x\cos x = x$

(ii) $\left(\frac{dy}{dx}\right)^2 + 6y^2 = 8x$

(iii) $\frac{d^2y}{dx^2} + 8\left(\frac{dy}{dx}\right) = -5x.$

Ordinary differential equation is a differential equation in which derivatives are involved with respect to a single independent variable.

Order of a differential equation is the order of the highest derivative appearing in the equation e.g.,

$$\frac{dy}{dx} + x \cos x = x$$

is a differential equation of first order. But the order of the differential equation (iii) as above is 2 as it involves the derivative of second order *viz.*,

$\frac{d^2 y}{dx^2}$.

Degree of a differential equation is the degree of the highest order derivative after the equation has been freed from the radical and fraction e.g.,

$\left(\frac{dy}{dx}\right)^2 + 6y^2 = 7x$ is 2 but degree of the diff. eq.

$\frac{d^2 y}{dx^2} = \sqrt{3 + \frac{dy}{dx}}$ is obtained after removing the radical and fractions. It is equivalent to

$\left(\frac{d^2 y}{dx^2}\right)^2 = 3 + \frac{dy}{dx}$ which is 2nd order and 2nd degree.

General solution is that solution of a differential equation which contains the number of arbitrary constants equal to the order of the differential equation.

Formation of differential equations: *Rule to form the differential equation from a given equation in x and y, containig arbitrary constants.*

1. Write down the given equation.
2. Differentiate w.r.t. x successively as many times as the number of arbitrary constants.
3. Eliminate the arbitrary constants from the equations of the above two steps.

The resulting equation is the required differential equation.

Different forms of differential equations

Type I. Differential equation $\frac{dy}{dx} = f(x)$.

we have $\frac{dy}{dx} = f(x)$, here we treat $\frac{dy}{dx}$ as $dy \div dx$

$\therefore\ dy = f(x)\, dx$

Integrating both sides, we get

$$\int dy = \int f(x)\, dx + c$$

where c is a constant of integration

$$\therefore\ y = \int f(x)\, dx + c$$

Example. *Solve* $\dfrac{dy}{dx} = \sin x - x$

$$dy = (\sin x - x)\, dx$$

Integrating both sides, we get

$$\int dy = \int (\sin x - x)\, dx$$

$$\int dy = \int \sin x\, dx - \int x\, dx + c$$

$$y = -\cos x - \frac{x^2}{2} + c$$

Type II. Variable Separable.

If in an equation, it is possible to get all the functions of x and dx to one side and all the functions of y and dy to the other, the variables are said to be separable.

Rule to solve an equation in which the variables are separable.

Consider the equation $\frac{dy}{dx} = XY$, where X is a function of x only and Y is a function of y only.

1. Given differential equation is $\frac{dy}{dx} = XY$
2. $\frac{dy}{Y} = X\,dx$ i.e., variables have been separated.
3. Integrating both sides $\int \frac{dy}{Y} = \int X\,dx + c$

   where c is an arbitrary constant.

Example. *Solve* $dy + xy\,dx = x\,dx$.

Sol. On separating the variables, we have

$$\frac{dy}{1-y} = x\,dx$$

Integrating both sides, we get

$$\int \frac{dy}{1-y} = \int x\,dx + c$$

$$\text{or} \quad -\log(1-y) = \frac{x^2}{2} + c$$

or $\frac{x^2}{2} + \log(1-y) = c.$

Type III. *Equations reducible to the form in which variables can be separated.*

Equation of the form $\frac{dy}{dx} = f(ax+by+c)$ can be reduced to the form in which the variables are separable.

Put $ax + by + c = z$, we have on differentiating w.r.t. x.

$$a + b\frac{dy}{dx} = \frac{dz}{dx}$$

or $$\frac{dy}{dx} = \frac{1}{b}\left(\frac{dz}{dx} - a\right)$$

$\therefore$ Given equation becomes

$$\frac{1}{b}\left(\frac{dz}{dx} - a\right) = f(z)$$

or $$\frac{dz}{dx} - a = b\,f(z) \text{ or } \frac{dz}{dx} = a + b\,f(z)$$

Separating the variables $\dfrac{dz}{a+bf(z)} = dx$

which can now be integrated.

Example. *Solve* $(x-y)^2 \dfrac{dy}{dx} = a^2$

Sol. Put $x-y=z$, then $1-\dfrac{dy}{dx} = \dfrac{dz}{dx}$ or $\dfrac{dy}{dx} = 1-\dfrac{dz}{dx}$

$\therefore$ Given equation becomes $z^2\left(1-\dfrac{dz}{dx}\right) = a^2$

or $\quad 1-\dfrac{dz}{dx} = \dfrac{a^2}{z^2}$

or $\quad \dfrac{dz}{dx} = 1-\dfrac{a^2}{z^2} = \dfrac{z^2-a^2}{z^2}$

Separating the variables

$$\frac{z^2}{z^2-a^2}\,dz = dx$$

or $\quad \left(1+\dfrac{a^2}{z^2-a^2}\right)dz = dx$

Integrating both sides,

$$z^2 + a^2 . \frac{1}{2a} \log \frac{z-a}{z+a} = x + c$$

or $$x - y + \frac{a}{2} \log \frac{x-y-a}{x-y+a} = x + c$$

or $$\frac{a}{2} \log \frac{x-y-a}{x-y+a} = y + c.$$

Type IV. *Equations of the form* $\frac{d^2y}{dx^2} = f(x)$.

We have, $\frac{d}{dx}\left(\frac{dy}{dx}\right) = f(x)$

Integrating w.r.t x, we get

$$\frac{dy}{dx} = \int f(x)\, dx + c_1$$

Let $\int f(x)\, dx = \phi(x)$ then $\frac{dy}{dx} = \phi(x) + c_1$

Again integrating w.r.t. x, we get

$$y = \int \phi(x)\, dx + c_1 x + c_2$$

Example. *Solve* $\frac{d^2y}{dx^2} = x.$

Sol. We have $\frac{d^2y}{dx^2} = x$ or $\frac{d}{dx}\left(\frac{dy}{dx}\right) = x$, Integrating

$$\int \frac{d}{dx}\left(\frac{dy}{dx}\right)dx = \int x\,dx + c_1 \text{ or } \frac{dy}{dx} = \frac{x^2}{2} + c_1$$

or $$\int \frac{dy}{dx}.dx = \int\left(\frac{x^2}{2} + c_1\right)dx + c_2$$

or $$y = \frac{x^3}{6} + c_1x + c_2.$$

First order linear equation with coefficients. A linear differential equation is that in which the dependent variables and its differential coefficients occur only in the first degree and are not multiplied together.

The standard form of linear differential equation of the first order is

$$\frac{dy}{dx} + Py = Q$$

P and Q are functions of x (and not of y) or constants.

Similarly, $$\frac{dx}{dy} + \mathrm{P}x = \mathrm{Q}$$

where P and Q are function of y only, is also called a linear differential equation of the first order.

Solution of differential equation $\frac{dy}{dx} + \mathrm{P}y = \mathrm{Q}$ is

$$ye^{\int P dx} = \int \mathrm{Q}.e^{\int P dx.} .dx + c$$

or y (integrating factor *i.e.* I.F,)

$$= \int \mathrm{Q}.(\mathrm{I.F.})\, dx + c$$

while evaluating the I.F. it is very useful to remember that

$$e^{\log f(x)} = f(x)$$

Illustration. *Solve* $\frac{dy}{dx} + y \tan x = \sec x.$

Sol. $\mathrm{P} = \tan x$ and $\mathrm{Q} = \sec x$ are functions of x only

$$\text{I.F.} = e^{\int P dx} = e^{\int \tan x \, dx} = e^{\log \sec x} = \sec x$$

$$y.(\text{I.F.}) = \int Q(\text{I.F.})\, dx + c$$

$$y \sec x = \int \sec x . \sec x \, dx + c$$

$$= \int \sec^2 x . dx + c$$

$$y \sec x = \tan x + c$$

$$y = \frac{\tan x}{\sec x} + \frac{c}{\sec x}$$

$$\therefore \quad y = \sin x + c \cos x$$

MATRICES AND DETERMINANTS

Definitions: A system of $m \times n$ numbers (real or complex) arranged in the form of an ordered set of m horizontal lines (called rows) and n vertical lines (called columns) is called an $m \times n$ matrix (read as m by n matrix).

The matrix of order $m \times n$ is written as

$$\begin{bmatrix} a_{11} & a_{12} & a_{13} & \ldots\ldots & a_{1j} & \ldots\ldots & a_{1n} \\ a_{21} & a_{22} & a_{23} & \ldots\ldots & a_{2j} & \ldots\ldots & a_{2n} \\ \ldots & \ldots & \ldots & \ldots\ldots & \ldots & & \ldots \\ a_{i1} & a_{i2} & a_{i3} & \ldots\ldots & a_{ij} & \ldots\ldots & a_{in} \\ \ldots & \ldots & \ldots & \ldots\ldots & \ldots & \ldots & \ldots \\ a_{m1} & a_{m2} & a_{m3} & & a_{mj} & & a_{mn} \end{bmatrix}$$

Notations: (*i*) Matrices are generally denoted by capital letters of the alphabet *viz.* A, B, C, ..., L, M, ..., X, Y ......

(*ii*) The elements are generally denoted by corresponding small letters *viz.* a_{ij}, b_{ij}, c_{ij} ..., l_{ij}, m_{ij}, ... x_{ij}, y_{ij}, ...

(*iii*) The following notations are used to enclose the elements which constitute a matrix.

[ ], ( ), || ||, { }

Types of matrices

(*i*) **Square matrix:** Any $n \times n$ matrix is called a square matrix of order n (or n-rowed matrix).

In this case, the number of row = the number of columns.

For example. (*a*) $\begin{bmatrix} 2 & 6 \\ 1 & 4 \end{bmatrix}$ is a square matrix of order 2.

(*b*) $\begin{bmatrix} 1 & 4 & 8 \\ 7 & 3 & 1 \\ 2 & 6 & 5 \end{bmatrix}$ is a square matrix of order 3.

(*ii*) **Rectangular matrix:** Any $m \times n$ matrix, where $m \neq n$ is called a rectangular matrix.

For example, $\begin{bmatrix} 2 & 7 & 8 \\ 3 & 2 & 6 \end{bmatrix}$ is a rectangular matrix.

(iii) **Row matrix:** Any $1 \times n$ matrix is called a row matrix. A row matrix has only one row.

For example. $[6 \quad 2 \quad 8]$ is a row matrix.

(iv) **Column matrix:** Any $m \times 1$ matrix is called a column matrix. A column matrix has only one column.

For example, $\begin{bmatrix} 3 \\ 2 \\ 7 \end{bmatrix}$ is a column matrix.

(v) **Principal diagonal:** The line along which the diagonal elements lie is said to be the principal diagonal. It is also called as leading diagonal or simply diagonal. Thus $a_{11}, a_{22} \ldots a_{ii}$ form the principal diagonal.

For example, If $A = \begin{bmatrix} 1 & 2 & 3 \\ 6 & -7 & 3 \\ -9 & 0 & 7 \end{bmatrix}$

then the diagonal elements are 1, –7, 7 and principal diagonal is the line on which these lies.

(*vi*) **Diagonal matrix:** A square matrix $A = [a_{ij}]$ is said to be diagonal matrix if $a_{ij} = 0$ when $i \neq j$.

Thus, it is a square matrix in which all the elements except the diagonal elements are zero.

For example, $\begin{bmatrix} 1 & 0 & 0 \\ 0 & 3 & 0 \\ 0 & 0 & 8 \end{bmatrix}$ is a diagonal matrix of order 3.

This is also denoted by symbol [1 3 8]

(*vii*) **Diagonal elements of a matrix:** An element a_{ij} of the square matrix $A = [a_{ij}]$ is said to be a diagonal element if $i = j$.

Thus $a_{11}, a_{22} \ldots a_{ij} \ldots$. are diagonal elements.

(*viii*) **Identity matrix:** A diagonal matrix is said to be an identity matrix if each of its diagonal elements is unity.

This is also known as unit matrix.

Thus $A = [a_{ij}]_{n \times n}$ is called an identity matrix. if (*i*) $a_{ij} = 0$ when $i \neq j$, (*ii*) $a_{ij} = 1$ when $i = j$

The identity matrix of order-n is usually denoted by I_n or simply I.

For example, $I_2 = \begin{bmatrix} 1 & 0 \\ 0 & 1 \end{bmatrix}$ $I_3 = \begin{bmatrix} 1 & 0 & 0 \\ 0 & 1 & 0 \\ 0 & 0 & 1 \end{bmatrix}$

(*ix*) **Null matrix:** A matrix is said to be a null matrix if each of its elements is zero.

This is also known as zero matrix.

The null matrix of the type $m \times n$ is denoted by $O_{m,n}$ or simply by O.

For example, $O_{2,3} = \begin{bmatrix} 0 & 0 & 0 \\ 0 & 0 & 0 \end{bmatrix}$

$$O_{3,3} = \begin{bmatrix} 0 & 0 & 0 \\ 0 & 0 & 0 \\ 0 & 0 & 0 \end{bmatrix}$$

Equality of matrices: Two matrices are said to be equal if and only if, (*i*) They are of the same order.

(*ii*) The elements in the corresponding positions are equal.

Note:

(*i*) If the two matrices A and B are equal, we can write them as A = B.

(*ii*) If the two matrices A and B are not equal, we write them as A ≠ B.

Operations on matrices: The following three operations on matrices are as follows:

(*i*) Addition of matrices

(*ii*) Multiplication of a martix by a scalar

(*iii*) Multiplication of matrices

Operation I. Addition of matrices. Let A and B be two matrices of the same type $m \times n$ then their sum (denoted by A + B) is defined as the matrix of the same type $m \times n$ obtained by adding the corresponding elements of A and B.

$$\text{Thus, if} \quad A = \begin{bmatrix} 2 & 3 & -7 \\ 8 & 2 & 9 \end{bmatrix}, \ B = \begin{bmatrix} 1 & -2 & 3 \\ 4 & 6 & 9 \end{bmatrix}$$

$$\text{Then } A + B = \begin{bmatrix} 2+1 & 3-2 & -7+3 \\ 8+4 & 2+6 & 9+9 \end{bmatrix} = \begin{bmatrix} 3 & 1 & -4 \\ 12 & 8 & 18 \end{bmatrix}$$

Note: Addition is defined only for matrices which are of the same type.

If two matrices are of the same type, they are said to be conformable for addition.

Difference of two matrices: Let A and B be two matrices of the same type $m \times n$, then their difference (denoted by A – B) is the sum of A and negative of B.

Thus, $A - B = A + (-B)$.

The difference A – B is obtained by subtracting from each element of A the corresponding element of B.

$$\text{Thus} \begin{bmatrix} 2 & 9 \\ 4 & 6 \end{bmatrix} - \begin{bmatrix} 1 & 0 \\ 3 & 1 \end{bmatrix} = \begin{bmatrix} 2-1 & 9-0 \\ 4-3 & 6-1 \end{bmatrix} = \begin{bmatrix} 1 & 9 \\ 1 & 5 \end{bmatrix}$$

Properties of Matrix Addition:

(*i*) **Matrix addition is commutative:** If A and B are any two matrices of the same type $m \times n$, then

$$A + B = B + A$$

(*ii*) **Matrix addition is associative:** If A, B and C are any three matrices of the same type $m \times n$, then

$$(A + B) + C = A + (B + C)$$

(*iii*) **Existence of additive identity:** If A and O are the matrices of the same type $m \times n$, then

A + O = O + A = A, where O is the null matrix.

Note: The null matrix O is known as additive identity and is unique.

(*iv*) **Existence of additive inverse:** If A and –A are the matrices of the same type $m \times n$ such that

$$A + (-A) = O = (-A) + A$$

Then –A is known as additive inverse of A.

(*v*) **Cancellation law:** If A, B and C are any three matrices of the same type $m \times n$.

Then $A + B = A + C \Rightarrow C = B$

[Left cancellation]

and $B + A = C + A \Rightarrow B = C$

[Right cancellation]

Operation II: Multiplication of a matrix by a scalar. Let A be any $m \times n$ matrix and K be any scalar. Then the $m \times n$ matrix obtained by multiplying every element of matrix A by K is known as scalar multiple of A by K and denoted by KA or AK.

Thus, if $A = \begin{bmatrix} 2 & 3 & 1 \\ 6 & 8 & 5 \end{bmatrix}$

Then,

$$8A = \begin{bmatrix} 8\times 2 & 8\times 3 & 8\times 1 \\ 8\times 6 & 8\times 8 & 8\times 5 \end{bmatrix} = \begin{bmatrix} 16 & 24 & 8 \\ 48 & 64 & 40 \end{bmatrix}$$

Operation III: Multiplication of Matrices.

If $A = \begin{bmatrix} a_{11} & a_{12} \\ a_{21} & a_{22} \end{bmatrix}$ and $B = \begin{bmatrix} b_{11} & b_{12} & b_{13} \\ b_{21} & b_{22} & b_{23} \end{bmatrix}$

Then the product of A and B (written as AB) is given by

$$AB = \begin{bmatrix} a_{11}b_{11} + a_{12}b_{21} & a_{11}b_{12} + a_{12}b_{22} & a_{11}b_{13} + a_{12}b_{23} \\ a_{21}b_{11} + a_{22}b_{21} & a_{21}b_{12} + a_{22}b_{22} & a_{21}b_{13} + a_{22}b_{23} \end{bmatrix}$$

$= C$ (say)

The entries in the product matrix are obtained as follows:

(*i*) The entry c_{11} in C, which lies in row 1 and column 1, is obtained by multiplying the elements of the first row of A by the corresponding elements of the first column of B. The sum of these products is c_{11}.

(*ii*) The entry c_{12} in C, which lies in row 1 and column 2 is obtained by multiplying the elements in the first row of A by the corresponding elements of second column of B. The sum of these product is c_{12}. Similarly other elements of C can be obtained.

In general. c_{ij} which lies in row i and column j is obtained by multiplying the elements of the ith row of A by the corresponding elements of jth column of B. The sum of these products is c_{ij}.

Note: The product AB of two matrices A and B is defined if and only if the number of columns of A = number of rows of B.

Two such matrices are said to be conformable for multiplication.

Properties of Matrix Multiplication:

(*i*) **Matrix multiplication is not commutative:** If A and B are any two matrices such that AB and BA are defined then AB ≠ BA. In fact, the following cases may arise:

1. AB may be defined but BA may not be defined.

2. BA may be defined but AB may not be defined.
3. AB and BA may be defined but $AB \neq BA$.
4. AB and BA may be defined and $AB = BA$.

(ii) **Matrix multiplication is associative:** If A, B and C are any three matrices such that A and B are conformable for the product AB , and B and C are conformable for the product BC, then

$$(AB)\,C = A\,(BC)$$

(iii) **Distributive law:** If A, B and C are any three matrices, then

$$A\,(B + C) = AB + AC$$

where the sum and products on both sides of the above relations are defined.

Similarly, $(B + C)\,A = BA + CA$.

(iv) **Cancellation law does not hold:** If A, B and C are matrices such that AB and AC are defined and also $AB = AC$, then it does not imply

$B = C$ or A cancels out on both sides.

Transpose of a Matrix: The matrix obtained from any given matrix A by interchanging the rows and columns of that matrix is called its transpose and is denoted by A′ or A^t.

For example. If $A=\begin{bmatrix}2 & 6 & 3\\ x & y & z\end{bmatrix}$ then $A'=\begin{bmatrix}2 & x\\ 6 & y\\ 3 & z\end{bmatrix}$

Determinants: Let us eliminate x and y from the equations

$$a_1x+b_1y=0$$

$$a_2x+b_2y=0$$

we get on elimination as

$$\frac{a_1}{b_1}=\frac{a_2}{b_2}\quad\left[\because \text{ each }=-\frac{y}{x}\right]$$

$$a_1b_2-a_2b_1$$

which in compact form can be written as

$$\begin{vmatrix}a_1 & b_1\\ a_2 & b_2\end{vmatrix}=0$$

In other words, $\begin{vmatrix}a_1 & b_1\\ a_2 & b_2\end{vmatrix}=a_1b_2-a_2b_1.$

The expression $\begin{vmatrix} a_1 & b_1 \\ a_2 & b_2 \end{vmatrix}$ which is a compact form of $a_1b_2 - a_2b_1$ is called a determinant of second order. The number a_1, b_1, a_2, b_2 are called the elements or constituents of the determinant.

Note:

(*i*) A determinant of second order is a function of $2^2 = 4$ elements arranged in two vertical lines called columns and two horizontal lines called rows in the form of a solid.

(*ii*) It has $2!$ $(= 2 \times 1 = 2)$ terms, half of them are positive and half are negative.

(*iii*) Each term in the expansion has one and only one element from each row and each column.

Let us eliminate x, y, z from the equations

$$a_1x + b_1y + c_1z = 0$$

$$a_2x + b_2y + c_2z = 0$$

$$a_3x + b_3y + c_3z = 0;$$

Solving the last two equations by cross multiplication

$$\frac{x}{b_2c_3 - b_3c_2} = \frac{y}{c_2a_3 - c_3a_2} = \frac{z}{a_2b_3 - a_3b_2} = k(\text{say})$$

$$x = k(b_2c_3 - b_3c_2);\ y = k(c_2a_3 - c_3a_2);\ z = k(a_2b_3 - a_3b_2).$$

Substituting these values of x, y and z in the first equation and cancelling k, we get

$$a_1(b_2c_3 - b_3c_2) + b_1(c_2a_3 - c_3a_2) + c_1(a_2b_3 - a_3b_2) = 0 \quad ...(i)$$

The L.H.S. expression of (*i*) can be written in compact form as $\begin{vmatrix} a_1 & b_1 & c_1 \\ a_2 & b_2 & c_2 \\ a_3 & b_3 & c_3 \end{vmatrix}$ which is called the determinant of third order and L.H.S. of (*i*) is called the expansion or the value of the determinant.

Note:

(*i*) A determinant of third order is a function of 3^2 (= 9) elements arranged in three vertical lines called columns and three horizontal lines called rows in the form of a solid square.

(*ii*) It has 3! (3 × 2 × 1 = 6) terms, half of them positive and half as negative.

(*iii*) Each term in the expansion has one and only one element from each row and each columns.

Expansion of second order determinant:

(+) (−)

$$\begin{vmatrix} a_1 & b_1 \\ a_2 & b_2 \end{vmatrix} = a_1b_2 - a_2b_1$$

sign of a product remains unchanged with downward arrow while it changes with upward arrow

(+) (−)

$$e.g. \begin{vmatrix} 3 & 6 \\ 2 & 1 \end{vmatrix} = 3 \times 1 - 6 \times 2 = 3 - 12 = 9$$

Expansion of third order determinant: The expansion of third order determinant is explained as

(*i*) Take the element of the first row and first column

$$e.g. \quad a_1 \begin{vmatrix} a_1 & b_1 & c_1 \\ a_2 & b_2 & c_2 \\ a_3 & b_3 & c_3 \end{vmatrix}$$

(*ii*) Reject all the elements of the row and the column to which a_1 belongs and form the determinant out of the remaining elements. We thus get the determinant of second order

$$\begin{vmatrix} b_2 & c_2 \\ b_3 & c_3 \end{vmatrix}$$

The determinant is called the minor of the element a_1, obtain the product of a_1 and its minor. It is written as

$$a_1 \begin{vmatrix} b_2 & c_2 \\ b_3 & c_3 \end{vmatrix}$$

(*iii*) Take the element of the first row and the second column *i.e.* b_1.

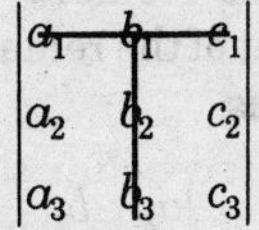

(*iv*) Reject all the elements of the row and the column to which b_1 belongs and form the determinant of the left out elements. You get the determinant

$$\begin{vmatrix} a_2 & c_2 \\ a_3 & c_3 \end{vmatrix}$$

It is called the minor of the element b_2. Obtain the product of the element b_1 and its minor, we get

$$b_1 \begin{vmatrix} a_2 & c_2 \\ a_3 & c_3 \end{vmatrix}$$

(*v*) Finally take the element of the first row and the third column *i.e.*, c_1.

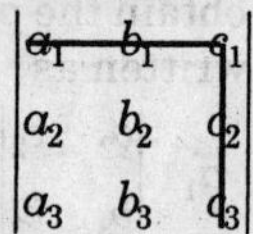

(*vi*) Reject all the elements of the row and the column to which c_1 belongs. Obtain the minor of the element c_1 by forming the determinant of the remaining element. The minor of c_1 is

$$\begin{vmatrix} a_2 & b_2 \\ a_3 & b_3 \end{vmatrix}$$

Find out the product of the element c_1 and its minor and get the product

$$c_1 \begin{vmatrix} a_2 & b_2 \\ a_3 & b_3 \end{vmatrix}$$

(*vii*) The product thus obtained are fixed with plus (+) and minus (–) signs alternately and then find their algebraic sum.

$$\begin{vmatrix} + & - & + \\ - & + & - \\ + & - & + \end{vmatrix}$$

The sum obtained is equal to the value of the original determinant of third order.

$$\begin{vmatrix} a_1 & b_1 & c_1 \\ a_2 & b_2 & c_2 \\ a_3 & b_3 & c_3 \end{vmatrix} = a_1\begin{vmatrix} b_2 & c_2 \\ b_3 & c_3 \end{vmatrix} - b_1\begin{vmatrix} a_2 & c_2 \\ a_3 & c_3 \end{vmatrix} + c_1\begin{vmatrix} a_2 & b_2 \\ a_3 & b_3 \end{vmatrix}$$

$$= a_1(b_2c_3 - b_3c_2) - b_1(a_2c_3 - a_3c_2) + c_1(a_2b_3 - a_3b_2)$$

Note:

(*i*) A determinant can be expanded by any row or by any column.

(*ii*) The sign before any term in the expansion is $(-1)^{i+j}$. For determinant of order 3, the signs are

$$\begin{vmatrix} + & - & + \\ - & + & - \\ + & - & + \end{vmatrix}$$

(*iii*) Expand from a row or column which contains maximum number of zeros.

Sarrus diagram for expansion of determinants of order 3: This is an easy method for evaluating a third order determinant but it is not applicable to determinants of order higher than 3.

Here repeat the first and second column to the right of the determinant

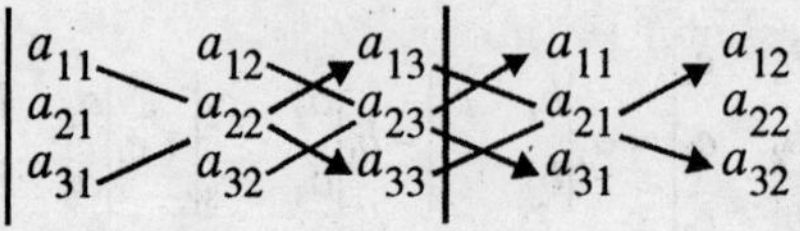

$$\left|\begin{array}{ccc} a_{11} & a_{12} & a_{13} \\ a_{21} & a_{22} & a_{23} \\ a_{31} & a_{32} & a_{33} \end{array}\right| \begin{array}{cc} a_{11} & a_{12} \\ a_{21} & a_{22} \\ a_{31} & a_{32} \end{array}$$

The term prefixed with +ve sign in the value of the determinant are those which correspond to the elements joined by continuous lines and the term prefixed with –ve sign are those which corresponds to the element joined by dotted lines in the above figure. The value of

$$\Delta = a_{11}a_{22}a_{33} + a_{12}a_{23}a_{31} + a_{13}a_{21}a_{32} - a_{31}a_{22}a_{13} - a_{32}a_{23}a_{11} - a_{33}a_{21}a_{12}$$

Minors and co-factors

Minor: The determinant obtained from a given determinant by omitting the row and the column in which a particular element lies is called the minor of that element.

Consider the determinant $\Delta = \begin{vmatrix} a_1 & b_1 & c_1 \\ a_2 & b_2 & c_2 \\ a_3 & b_3 & c_3 \end{vmatrix}$

To find the minor of c_2. c_2 occurs in the second row and third column. Omitting the second row and third column, the minor of c_2 is

$$\begin{vmatrix} a_1 & b_1 \\ a_3 & b_3 \end{vmatrix}$$

Co-factor: Co-factors are minors with proper sign and are generally denoted by the corresponding capital letters.

The co-factor of a particular element is $(-1)^{i+j}$ times the determinant obtained from the given determinants by omitting the row and a column in which it lies where i is the number of row and j the number of column, in which that element lies.

e.g., C_2 = co-factor of c_2 in Δ

$$= (-1)^{2+3} \begin{vmatrix} a_1 & b_1 \\ a_3 & b_3 \end{vmatrix} = -\begin{vmatrix} a_1 & b_1 \\ a_3 & b_3 \end{vmatrix} \left[\because (-1)^5 = -1\right]$$

Properties of determinant:

Property I: If each element of a row or a column of a determinant is zero, then the value of the determinant is zero

$$i.e., \begin{vmatrix} 0 & 0 & 0 \\ a_2 & b_2 & c_2 \\ a_3 & b_3 & c_3 \end{vmatrix} = 0 \text{ and } \begin{vmatrix} 0 & b_1 & c_1 \\ 0 & b_2 & c_2 \\ 0 & b_3 & c_3 \end{vmatrix} = 0$$

Property II: The value of a determinant is not altered by changing its rows into columns and columns into rows.

Let $\Delta = \begin{vmatrix} a_1 & b_1 & c_1 \\ a_2 & b_2 & c_2 \\ a_3 & b_3 & c_3 \end{vmatrix}$ be the given determinant.

Let Δ' be the determinant obtained from Δ by changing its rows into columns and columns into rows.

$$\Delta' = \begin{vmatrix} a_1 & a_2 & a_3 \\ b_1 & b_2 & b_3 \\ c_1 & c_2 & c_3 \end{vmatrix}$$

Then, $\Delta = \Delta'$

Property III: If two adjacent rows or columns of a determinant are interchanged, the determinant changes in sign but its numerical value is unaltered.

$$\text{Let } \Delta = \begin{vmatrix} a_1 & b_1 & c_1 \\ a_2 & b_2 & c_2 \\ a_3 & b_3 & c_3 \end{vmatrix}$$

Let Δ' be the determinant obtained from Δ by interchanging its 1st and 2nd rows.

$$\Delta' = \begin{vmatrix} b_1 & a_1 & c_1 \\ b_2 & a_2 & c_2 \\ b_3 & a_3 & c_3 \end{vmatrix}$$

Then, $\Delta' = -\Delta$

Property IV: If two rows or columns of a determinant are identical (same) then the value of the determinant is zero.

$$\Delta = \begin{vmatrix} a_1 & a_1 & c_1 \\ a_2 & a_2 & c_2 \\ a_3 & a_3 & c_3 \end{vmatrix} = 0$$

[Here, 1st and 2nd column are identical]

Property V: If each element of a row or column of a determinant be multiplied by the same factor, then the determinant is multiplied by that factor.

Let $$\Delta = \begin{vmatrix} a_1 & b_1 & c_1 \\ a_2 & b_2 & c_2 \\ a_3 & b_3 & c_3 \end{vmatrix}$$

Let Δ' be the determinant obtained from Δ by multiplying the elements of the first column by a constant k.

$$\Delta' = \begin{vmatrix} ka_1 & b_1 & c_1 \\ ka_2 & b_2 & c_2 \\ ka_3 & b_3 & c_3 \end{vmatrix}$$

Then, $\Delta' = k\,\Delta$

Property VI: If each element of a row or column of a determinant can be expressed as the sum of two (or more) terms, then the determinant can also be written as the sum of two (or more) determinants.

$$i.e., \begin{vmatrix} a_1+\alpha_1 & b_1 & c_1 \\ a_2+\alpha_2 & b_2 & c_2 \\ a_3+\alpha_3 & b_3 & c_3 \end{vmatrix} = \begin{vmatrix} a_1 & b_1 & c_1 \\ a_2 & b_2 & c_2 \\ a_3 & b_3 & c_3 \end{vmatrix} + \begin{vmatrix} \alpha_1 & b_1 & c_1 \\ \alpha_2 & b_2 & c_2 \\ \alpha_3 & b_3 & c_3 \end{vmatrix}$$

Property VII: If each element of a row (or column) of a determinant be added or subtracted the equimultiples of the corresponding elements of one or more rows (or columns), the determinant remains unaltered.

$$\text{Let } \Delta = \begin{vmatrix} a_1 & b_1 & c_1 \\ a_2 & b_2 & c_2 \\ a_3 & b_3 & c_3 \end{vmatrix}$$

Let Δ' be the determinant obtained from Δ by adding p times the elements of the second column and subtracting q times the elements of the third column from the first column.

$$\Delta' = \begin{vmatrix} a_1+pb_1-qc_1 & b_1 & c_1 \\ a_2+pb_2-qc_2 & b_2 & c_2 \\ a_3+pb_3-qc_3 & b_3 & c_3 \end{vmatrix}$$

Then, $\Delta' = \Delta$

A consistent system of non-homogeneous linear equations: If a system of linear equations has a solution, we say that the system is consistent. A consistent system of linear equations may have a unique solution or an infinite number of solutions. e.g.,

(A) Let $x+y=3$ and $x-y=1$ be a system of two non-homogeneous linear equations which on solving gives the value of $x=2, y=1$ and therefore these equations are consistent. The solution $x=2, y=1$ is a unique solution.

(B) Let $x+y=3$ and $3x+3y=9$ be another system of linear equations. These equations are satisfied by $x=0, y=3$; $x=1, y=2$ etc. There are infinite solutions of the given equation.

Inconsistent system: If a system of linear equations has no solution (*i.e.*, if there exists no common values of x, y or x, y, z satisfying the given equations we say the system of linear equations is inconsistent. For example $x+y=2$ and $x+y=3$ is a system of inconsistent equations because there exists no values of x and y which satisfy both the equations. [From the two equations equating the two values of $x+y$, we get $2=3$ which is not possible].

Crammer's Rule to solve linear equations by determinants: Consider three linear equations in three unknowns.

$$a_1x + b_1y + c_1z = d_1 \quad ...(i)$$

$$a_2x + b_2y + c_2z = d_2 \quad ...(ii)$$

$$a_3x + b_3y + c_3z = d_3 \quad ...(iii)$$

Let D = the determinant of the coefficient of x, y, z in (i), (ii) and (iii)

$$= \begin{vmatrix} a_1 & b_1 & c_1 \\ a_2 & b_2 & c_2 \\ a_3 & b_3 & c_3 \end{vmatrix}$$

Replacing the elements of first column by d_1, d_2, d_3.

Let $$D_1 = \begin{vmatrix} d_1 & b_1 & c_1 \\ d_2 & b_2 & c_2 \\ d_3 & b_3 & c_3 \end{vmatrix}$$

Replacing the elements of second column by d_1, d_2, d_3.

Let $$D_2 = \begin{vmatrix} a_1 & d_1 & c_1 \\ a_2 & d_2 & c_2 \\ a_3 & d_3 & c_3 \end{vmatrix}$$

Replacing the elements of third column by d_1, d_2, d_3.

Let
$$D_3 = \begin{vmatrix} a_1 & b_1 & d_1 \\ a_2 & b_2 & d_2 \\ a_3 & b_3 & d_3 \end{vmatrix}$$

Substituting the values of d_1, d_2, d_3 from (*i*), (*ii*) and (*iii*) in (*iv*)

$$D_1 = \begin{vmatrix} d_1 & b_1 & c_1 \\ d_2 & b_2 & c_2 \\ d_3 & b_3 & c_3 \end{vmatrix}$$

We have

$$D_1 = \begin{vmatrix} a_1x + b_1y + c_1z & b_1 & c_1 \\ a_2x + b_2y + c_2z & b_2 & c_2 \\ a_3x + b_3y + c_3z & b_3 & c_3 \end{vmatrix}$$

$$= \begin{vmatrix} a_1x & b_1 & c_1 \\ a_2x & b_2 & c_2 \\ a_3x & b_3 & c_3 \end{vmatrix} + \begin{vmatrix} b_1y & b_1 & c_1 \\ b_2y & b_2 & c_2 \\ b_3y & b_3 & c_3 \end{vmatrix} + \begin{vmatrix} c_1z & b_1 & c_1 \\ c_2z & b_2 & c_2 \\ c_3z & b_3 & c_3 \end{vmatrix}$$

$$= x\begin{vmatrix} a_1 & b_1 & c_1 \\ a_2 & b_2 & c_2 \\ a_3 & b_3 & c_3 \end{vmatrix} + y\begin{vmatrix} b_1 & b_1 & c_1 \\ b_2 & b_2 & c_2 \\ b_3 & b_3 & c_3 \end{vmatrix} + z\begin{vmatrix} c_1 & b_1 & c_1 \\ c_2 & b_2 & c_2 \\ c_3 & b_3 & c_3 \end{vmatrix}$$

(Two columns are same) (Two columns are same)

$$= x(\text{D}) + y(0) + z(0) = x\text{D}$$

or $x\text{D} = \text{D}_1$ ...(*v*)

Similarly $y\text{D} = \text{D}_2$ ...(*vi*)

and $z\text{D} = \text{D}_3$ ...(*vii*)

If $\text{D} \neq 0$, then $x = \dfrac{\text{D}_1}{\text{D}}, y = \dfrac{\text{D}_2}{\text{D}}, z = \dfrac{\text{D}_3}{\text{D}}$

(unique solution)

Note 1: If $\text{D} = 0, \text{D}_1 = 0, \text{D}_2 = 0, \text{D}_3 = 0$ then (*v*), (*vi*) and (*vii*) become $x(0) = 0, y(0) = 0, z(0) = 0$. These equations are satisfied by all values of x, y and z (an infinite solutions).

Note 2: If $\text{D} = 0$ but at least one of $\text{D}_1, \text{D}_2, \text{D}_3$ is non-zero then the value of at least one of x, y, z tends of infinity and therefore the solution does not exist. Hence, equations are inconsistents.

Singular and non-singular matrix: A square matrix A is said to be singular if $|A| = 0$ and non-singular if $|A| \neq 0$.

Adjoint of a square matrix: Let $A = [a_{ij}]$ be a square matrix of order n. Then the adjoint of the matrix A is the matrix $B = [b_{ij}]$ where $b_{ij} = a_{ij}$ (where A_{ji} is the cofactor of a_{ij} in the determinant A).

The adjoint of a martix A is also called **Adjugate** of a matrix and is denoted by **adj. A.**

Example : Calculate the adjoint of A where

$$A = \begin{vmatrix} 1 & 1 & 1 \\ 1 & 2 & -3 \\ 2 & -1 & 3 \end{vmatrix}$$

Sol : Let the adj $A = B = [b_{ij}]$

$$b_{11} = \text{the cofactor of } a_{11} \text{ in } |A| = \begin{vmatrix} 2 & -3 \\ -1 & 3 \end{vmatrix} = 3$$

$$b_{21} = \text{the cofactor of } a_{12} \text{ in } |A| = -\begin{vmatrix} 1 & -3 \\ 2 & 3 \end{vmatrix} = -9$$

$$b_{31} = \text{the cofactor of } a_{13} \text{ in } |A| = \begin{vmatrix} 1 & 2 \\ 2 & -1 \end{vmatrix} = -5$$

$$b_{12} = \text{the cofactor of } a_{21} \text{ in } |A| = \begin{vmatrix} 1 & 1 \\ -1 & 3 \end{vmatrix} = -4$$

$$b_{22} = \text{the cofactor of } a_{22} \text{ in } |A| = \begin{vmatrix} 1 & 1 \\ 2 & 3 \end{vmatrix} = 1$$

$$b_{32} = \text{the cofactor of } a_{23} \text{ in } |A| = -\begin{vmatrix} 1 & 1 \\ 2 & -1 \end{vmatrix} = 3$$

$$b_{13} = \text{the cofacor of } a_{31} \text{ in } |A| = \begin{vmatrix} 1 & 1 \\ 2 & -3 \end{vmatrix} = -5$$

$$b_{23} = \text{the cofactor of } a_{32} \text{ in } |A| = -\begin{vmatrix} 1 & 1 \\ 1 & -3 \end{vmatrix} = 4$$

$$b_{33} = \text{the cofactor of } a_{33} \text{ in } |A| = \begin{vmatrix} 1 & 1 \\ 1 & 2 \end{vmatrix} = 1$$

$$\text{adj. } A = B = \begin{vmatrix} b_{11} & b_{12} & b_{13} \\ b_{21} & b_{22} & b_{23} \\ b_{31} & b_{32} & b_{33} \end{vmatrix} = \begin{vmatrix} 3 & -4 & 5 \\ -9 & 1 & 4 \\ -5 & 3 & 1 \end{vmatrix}$$

Theorem: If A is an n-rowed square matrix, then A (adj. A) = (adj. A) A = $|A|\ I_n$; where I_n is a n-rowed unit matrix.

Inverse of a square matrix: Let A be a square matrix of order $n \times n$. If there exists another square matrix B of order $n \times n$ such that $AB = BA = I_n$, then B is called inverse or reciprocal of the square matrix A and the matrix A is known as invertible. It is denoted by A^{-1}. So that $AA^{-1} = A^{-1}A = I_n$; obviously $(A^{-1})^{-1} = A$.

Note 1: A rectangular matrix does not possess inverse.

Note 2: If B is the inverse of A, surely A is also inverse of B.

Theorem 1: Inverse of every square matrix if it exists is unique.

Theorem 2: The necessary and sufficient condition for a square matrix A to possess an inverse is that it must be non-singular.

Note: Singular matrices cannot have inverses.

Cor.1. $$A^{-1} = \frac{\text{adj. A}}{|A|}$$

Cor. 2. If $$A = \begin{vmatrix} a_{11} & a_{12} & a_{13} \\ a_{21} & a_{22} & a_{23} \\ a_{31} & a_{32} & a_{33} \end{vmatrix}$$

$$\text{then } A^{-1} = \frac{1}{|A|}\begin{vmatrix} A_{11} & A_{21} & A_{31} \\ A_{12} & A_{22} & A_{32} \\ A_{13} & A_{23} & A_{33} \end{vmatrix}$$

where A_{ij} is the cofactor of a_{ij} in $|A|$, for $i, j = 1, 2, 3$

Example: Compute the inverse of the matrix

$$\begin{bmatrix} 2 & 0 & -1 \\ 5 & 1 & 0 \\ 0 & 1 & 3 \end{bmatrix}$$

Sol: Let the given matrix be denoted by A, then

$$|A| = \begin{vmatrix} 2 & 0 & -1 \\ 5 & 1 & 0 \\ 0 & 1 & 3 \end{vmatrix} = 6 - 5 = 1 \neq 0$$

Thus A^{-1} exists. Let us calculate the adj. A.

The cofactors of the elements of the first row in $|A|$ are

$$\begin{vmatrix} 1 & 0 \\ 1 & 3 \end{vmatrix}, -\begin{vmatrix} 5 & 0 \\ 0 & 3 \end{vmatrix}, \begin{vmatrix} 5 & 1 \\ 0 & 1 \end{vmatrix} \quad i.e., 3, -15, 5$$

The cofactors of the elements of the second row in $|A|$ are

$$-\begin{vmatrix} 0 & -1 \\ 1 & 3 \end{vmatrix}, \begin{vmatrix} 2 & -1 \\ 0 & 3 \end{vmatrix}, -\begin{vmatrix} 2 & 0 \\ 5 & 1 \end{vmatrix} \text{ i.e., } -1, 6, -2$$

The cofactors of the elements of the 3rd row in $|A|$ are

$$\begin{vmatrix} 0 & -1 \\ 1 & 0 \end{vmatrix}, -\begin{vmatrix} 2 & -1 \\ 5 & 0 \end{vmatrix}, \begin{vmatrix} 2 & 0 \\ 5 & 1 \end{vmatrix} \text{ i.e., } 1, -5, 2.$$

$$\text{Adj. } A = \begin{bmatrix} 3 & -1 & 1 \\ -15 & 6 & -5 \\ 5 & -2 & 2 \end{bmatrix}$$

$$A^{-1} = \frac{\text{Adj. A}}{|A|} = \begin{vmatrix} 3 & -1 & 1 \\ -15 & 6 & -5 \\ 5 & -2 & 2 \end{vmatrix}$$

Working Rule for solving the linear equations:

$a_1x + b_1y = c_1$

$a_2x + b_2y = c_2$

Step I: Write down the given system of equations in a single matrix equation AX = B.

where $A = \begin{bmatrix} a_1 & b_1 \\ a_2 & b_2 \end{bmatrix}, X = \begin{bmatrix} x \\ y \end{bmatrix}, B = \begin{bmatrix} c_1 \\ c_2 \end{bmatrix}.$

Step II: Evaluate $|A|$, i.e., find $|A| = a_1b_2 - a_2b_1$

Case I: When $|A| \neq 0$.

Step III: If the matix A is non-singular i.e., $|A| \neq 0$ then A^{-1} exists and find A^{-1}.

Step IV: Pre-multiply both sides of the equation $AX = B$ by A^{-1} so that

We have $A^{-1}AX = A^{-1}B \Rightarrow X = A^{-1}B$

$[\because AA^{-1} = I]$

Step V: Equate the two matrices and get the values of x and y.

Case. II: When $|A| = 0$. In this case A^{-1} does not exist.

There are two possibilities.

(a) When $\dfrac{a_1}{a_2} = \dfrac{b_1}{b_2} \neq \dfrac{c_1}{c_2}$

In this case the given system of equation are *inconsistant*.

(b) When $\dfrac{a_1}{a_2} = \dfrac{b_1}{b_2} = \dfrac{c_1}{c_2}$

In this case the equations are consistent and have an infinite solutions.

25

MENSURATION

1. Rectangle:

(a) Area of a rectangle = length × breadth

(b) Perimeter of a rectangle = 2(length + breadth)

(c) Diagonal of a rectangle $= \sqrt{(\text{length})^2 + (\text{Breadth})^2}$

2. Square:

(a) Area of a square = $(\text{side})^2$

(b) Perimeter of a square = 4 × side

(c) Diagonal of a square = $\sqrt{2}$ × side

3. Circle:

(a) Area of a circle = $\pi \times (\text{radius})^2$

(b) Circumference of a circle = $2\pi \times (\text{radius})$

(c) Radius $= \dfrac{\text{Diameter}}{2}$ or

Diameter = 2 × Radius

4. Triangle:

(a) Area of a triangle = $\sqrt{s(s-a)(s-b)(s-c)}$

where $s = \dfrac{a+b+c}{2}$

(b) Area of a right angled triangle

$= \dfrac{1}{2} \times$ base (b) $\times$ height (h)

(c) Area of an equilateral triangle

$= \dfrac{\sqrt{3}}{4} \times (\text{side})^2$

5. Quadrilateral:

(a) Area of a quadrilateral = $\dfrac{1}{2}$ × diagonal × sum of the offsets

Here AC = diagonal, BF and DE are offsets

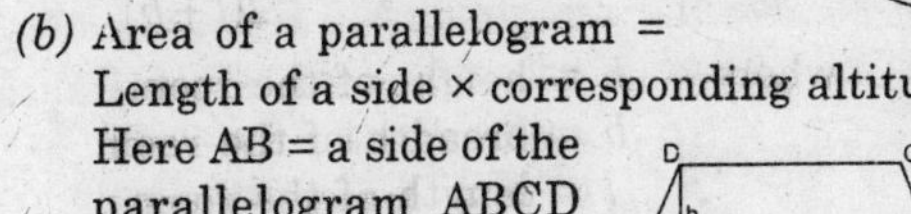

(b) Area of a parallelogram = Length of a side × corresponding altitude

Here AB = a side of the parallelogram ABCD and h = corresponding altitude

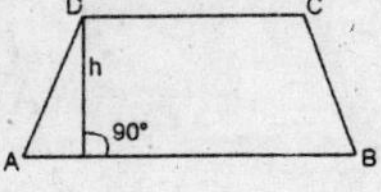

(*c*) Area of a rhombus $= \frac{1}{2} \times$ product of the diagonals

Here AC, BD = diagonals
and AB = BC = CD = DA

(*d*) Area of a trapezium

$= \frac{1}{2} \times$ (sum of the parallel sides) $\times$ distance between them

Here AB, CD = parallel sides, CE = distance between the parallel sides.

6. Four Walls:

(*a*) Area of the four walls (A) = $2 \times h(l + b)$

(*b*) Height of the room $(h) = \dfrac{A}{2(l+b)}$

where h = height of the room
b = breadth of the room
l = length of the room
A = area of the four walls